Praise for

Running Ransom Road

"In *Running Ransom Road,* Caleb Daniloff reminds us that running is far more than a sport, as a marathon can be far more than just a 26.2-mile foot race. As he skillfully demonstrates in his words, running changes lives and in many cases can even save lives. After years of addiction, anger, and confusion, the run became just the healing agent Daniloff needed. A joy to read."

— **Dave McGillivray, Race Director, B.A.A. Boston Marathon**

"Caleb Daniloff is not a wannabe distance god–type runner. But after reading his book, *Running Ransom Road,* I can see the terrific effort Caleb has made. A world class effort. Read his book, and you will get it about running. Like I've always said, this is more than sport. No sport is more real, more demanding than marathoning. His story is sad for a long time, but bright in the unfolding."

— **Bill Rodgers, four-time Boston Marathon and New York City Marathon winner and author of *Marathon Man***

"Distinctive and exciting."

— **Rebecca Shapiro, *Columbia Magazine***

Running Ransom Road

Running
Ransom Road

CONFRONTING THE PAST,
ONE MARATHON AT A TIME

Caleb Daniloff

Mariner Books
Houghton Mifflin Harcourt
BOSTON · NEW YORK

First Mariner Books edition 2013

Copyright © 2012 by Caleb Daniloff

www.hmhbooks.com

Library of Congress Cataloging-in-Publication Data
Daniloff, Caleb.
Running Ransom road : confronting the past, one marathon at a time /
Caleb Daniloff.
p. cm.
Includes index.
ISBN 978-0-547-45005-6   ISBN 978-0-544-10543-0 (pbk.)
1. Marathon running. I. Title.
GV1065.D36 2012
796.42'52—dc23

Book design by Brian Moore
Maps by Paul Lobue

Printed in the United States of America
DOC 10 9 8 7 6 5 4 3 2 1

The author is grateful for permission to reprint "oh yes," from *War All the Time:
Poems 1981–1984* by Charles Bukowski. Reprinted by permission
of HarperCollins Publishers.

The excerpt from "It Was Easy as Pie," by Sarah Pileggi, is reprinted by
permission of *Sports Illustrated,* from the December 3, 1979, issue.

To Chris and Shea, never a question

Author's Note

While some names have been changed, I have done my level best to faithfully re-create events from my past. Some people may remember the same circumstances differently. We all see things by our own light. This is mine.

Contents

Out beyond all ideas of rightness and wrongness,
there is a field. I'll meet you there.

— RUMI

oh yes

there are worse things than
being alone
but it often takes decades
to realize this
and most often
when you do
it's too late
and there's nothing worse
than
too late.

— CHARLES BUKOWSKI

Prologue

Longfellow Bridge Loop

March 2008

CAMBRIDGE, MASSACHUSETTS

It's still dark out. Rain smears my bedroom window and pours off the streetlamps into icy puddles below. Drops pelt the sidewalks, splashing furiously off sheets of water as if the world has been set to boil. It's been several days since my last run and the extra weight I'm feeling is more than last night's pepperoni pizza. I throw back the covers and tug on socks, windpants, and a jacket. I hear that shiftless part of me whisper, This is stupid; just go back to bed. *The voice that had gotten me into trouble over the years—that assured me I had room for another drink, another party to crash, could see straight enough to drive—is still persuasive, especially when the mercury reads a raw 34 degrees.*

For fifteen years, from the ages of fourteen to twenty-nine, I often found myself drunk or hung-over, usually both. In college, I earned the nickname "Asshole" and proudly answered. Drunkenness was my calling. I worked hard at it—at the bars, on the streets, behind the wheel. Pass me that after-shave, goddamn it. No one can black out like me. *Needless to say, the only part of me that ran back then was my mouth, whether I was locked in a shouting match with a girlfriend, begging a couple of bucks for a shot, or pleading with a store clerk who'd caught me stuffing a bottle of wine down my pants. And those were the good years.*

It's been almost a decade since I last wiped Budweiser foam from my lips. I don't wake up hung-over anymore, but I do sometimes wake up haunted—by who I used to be, by the people I've done wrong. On the days I don't run, it's worse. I'm filled with a different kind of thirst, a need to move between places—across bridges, over water, over city lines. The nastier the conditions, the better: lightning storms, ice-covered sidewalks, predawn country roads during hunting season. The hard work of the run fortifies my will. I move through this so I can move through that. One foot in front of the other. One run at a time.

By the time I reach the Charles River, my socks and gloves are sponges, my pants shining with water, my face a mask of rain. A lone minivan rolls down Memorial Drive, spastic wipers manically clearing the windshield. The river gulps at the banks. Even the Canada geese have taken shelter under the trees. Without my glasses, the downtown skyline is a gray smear. Office lights or stars, I can't tell.

Over the years, I'd been sent to Alcoholics Anonymous by boarding school administrators and later the courts, but I was never able to connect with the religious-tinged language, the intense gratefulness, the slow, plodding work, the surrender. Everyone seemed older, more grizzled, more committed. I earned myself a thirty-day chip so my friends and parents couldn't call my out-of-control behavior out-of-control. That little blue disk carved with the double As only became a license to party harder. Drinking remained the central fact in my life, its only rhythm.

But after I got sober for real nine years later, the anxieties and insecurities I'd tried to cover up with booze still remained. Only they'd morphed as if exposed to radiation. Shyness had become panic. Self-doubt was self-loathing. I didn't know how to be friendly. I was impatient, but my mind was as nimble as a tree sloth. I didn't get jokes, only perceived slights. I was forgetful, quick to tremble, a little unsteady on my feet. I'd grown so used to opening my eyes in terror, confusion, and shame over the years that I was sure I'd worn permanent grooves into my brain—nightmares still jetted from my brain's limbic system to the screen behind my eyeballs, smooth bullets fitted into an even smoother chamber. I could taste the absence of vodka in my orange juice.

Harboring memories of disconnect, I chose not to return to the hard-eyed groups and those church basements and instead to gut it out alone. But as those first sober years passed, the details of my offenses, the faces and voices of those I hurt, started to dissolve. The guilt calcified. Over time, the past became a hard lump in my throat, a walnut I had trouble swallowing as I pushed a lawn mower across the grass of my first home, cheered my stepdaughter on the soccer field, made small talk with church-going parents at show-and-tell evenings at the elementary school. I was reserved, quiet, the last person they'd picture vomiting out a car window in a crowded city square, then cracking another beer and toasting the shocked faces.

But when I took up running, I found not only a new central pattern to my life, but a forum in which to confront myself. Have I avoided AA because I'm afraid to say, "I'm an alcoholic" in a roomful of strangers? Was it cowardly to write apology letters rather than look people in the eye? Can a former drunk ever be truly happy again?

As I pound across the puddle-filled Longfellow Bridge toward

Boston, Beacon Hill shrouded in mist, I grapple with the ghost of me and how it fits within my current form, the one clocking nine-minute miles in motion-control running shoes, with the wife and teenage stepdaughter, the mortgage and aging parents. At the span's last stone bastion, I hop down the slick concrete steps that lead to the pedestrian walkway and wind back down to the river.

On the Esplanade, an empty sleeping bag hangs off a bench, a pair of drenched tennis shoes parked beneath, along with pieces of soggy cardboard, detritus of a homeless life. Often, when I run past the bottle-strewn camps on the riverbank, guys nursing bagged beers and staring at the water, nowhere to go, nowhere to be, I wonder about the fateful combination of decisions and life events that has separated them from the centrifugal forces that keep the rest of us pinned in place. I usually wave as I pass.

In the back of my mind, I harbor a similar wonder about my life. How does a shy, athletic child from a stable family turn into an obnoxious, chain-smoking drunk? I wish I could point to some horrific trauma—an abusive babysitter, a terrible car crash—that I've buried so deep it's become somebody else's problem in China. My parents weren't big drinkers, though alcoholism had cast shadows on both sides of the family. There was no divorce, no abuse; we lived in safe neighborhoods. Had I taken my teenage itch to fit in too far? Maybe it had something to do with childhood bed-wetting, which burned me with shame and taught me to lie. Depression ran in my dad's family.

Churning through the wet, empty darkness, past the Community Boating boathouse and the half dome of the Hatch Shell, I draw lessons from the road: The hard, cracked surface of the sidewalk reminds me of the facts in my life that have to be accepted—that I've cheated on girlfriends, abandoned friends, spewed cruel words. The lung-squeezing hills tell me about the pain that precedes reward. The weather—the rain, the snow, the ice—is the little stuff life throws at you. And sometimes it's the little stuff, a broken shoelace, the finger from a pissed driver, a botched fact in one of your articles, that can cartwheel you back toward your dark places.

I don't know whether I'll ever fully calm the waters of my past, but the steady drumbeat of my feet on the ground and my arms sawing through the air helps. For an hour at a time, I am enduring, rebuilding. The harder I push and the farther I go, the more it hurts and the better it feels. Softened, I'm able to pry my body from the space it occupies, and in that cracked-open territory is a place where I can touch my various facsimiles: the scrawny sixth-grader stripping his piss-soaked bed for the fifth morning in a row, the shy thirteen-year-old eating lunch in the bathroom stall of a new school, the cocky seventeen-year-old drunk on his fifteen minutes of fame, the blurry-eyed twenty-three-year-old speeding down the highway on blind, angry tires. And it's where I can try to forgive the thirty-eight-year-old with the shin splints and the sweat stinging his eyes.

Through the curtains of rain, I come across another runner by one of the stone footbridges. A middle-aged man with a slight paunch, drenched gray sweatshirt, and knit cap dripping with water. He has a mustache and thick eyebrows. I wonder what's brought him out in this foul stew. We raise hands as we pass and look into the other's eyes. Is he confronting something, too—a medical diagnosis? A divorce? A death? We recognize each other, that a particular kind of work is being done out here. Then he's gone, footfalls smacking, receding. But he stays in my mind as I splash beneath the Boston University Bridge. Both of us alone out here, a community of two chasing things that will never be caught. No longer do I run from my demons, but run with them. We pace each other, the past and me. And some days, I go faster.

Running Ransom Road

Boston

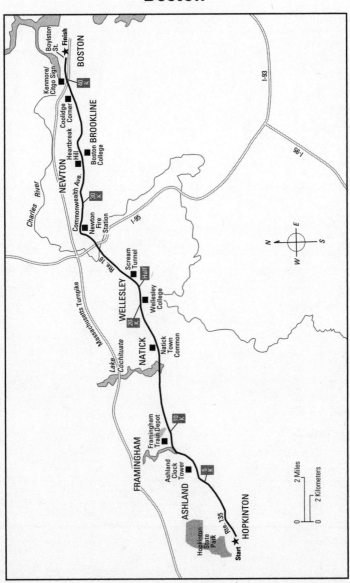

1

113th Boston Marathon

Monday, April 20, 2009

I FELT JITTERY APPROACHING the mouth of Route 135 East in Hopkinton. It was a crisp February morning. The sidewalks were empty. Snow was spitting as my eyes teased out the faded unicorn logo of the Boston Athletic Association painted in the road — the twisted horn, the flared nostrils, the proud toss of the neck. Cars rumbled by, shaking road salt from their undercarriages as they streamed across running's most venerable starting line. I bent and retightened my laces, then adjusted my fuel belt, a bandolier of plastic containers sloshing with liquid the color of wiper fluid. I toed the line for a few seconds, then crossed over, half expecting the other side to feel different as if the stripe were a palpable separation between Yesterday and Today. But my mind snapped back to the task at hand. I was now scuffing down the hallowed grounds of the historic Boston Marathon, pores wide open, ready to mainline some serious running mojo. So what if the gun was still two months off?

These were my first steps on any marathon course. That it was the world's oldest continuous 26.2-miler only made it more daunting. The road sloped, flattened, then dropped again. I paid attention to the downhills. Take them too fast, I was told, and you could find yourself later with anvils for thighs. Restraint had to be part of the plan, mind over adrenaline, a tall order given the electricity sure to be crackling on race day. Starting in the small suburban burg of Hopkinton and ending in Boston's bustling Copley Square, this eight-town course had been the route since 1927 and I wanted to drink it all in, pound it like the six-packs that used to fizz my brain. I passed a few nurseries, a Christmas tree farm, a park, and a horse-riding ring. So far, so good. I wondered if I'd see plaques along the way, statues, bronze mile markers, hear the harps of angels.

The air was cold and sharp, scraping the bottom of my lungs, my fingertips tingling in my gloves. The breeze whispered against my cheeks and sweat began forming above my lip like puberty. This was my first training run away from the flat, leafy paths along the Charles River. My feet pounded the ground, absorbing the unfamiliar road, a handshake of sorts. My heart was still unsettled in my rib cage, teetering in that moment between adrenaline flow and the emptiness of pace. It was in that moment when my run might go in any direction, when it was deciding what it wanted to be. I could feel my brain powering down, my mind humming to life.

Over the next few miles, the landscape morphed, revealing a Dunkin' Donuts, an automotive repair shop, a paint store, a commuter rail parking lot. Ranch houses and modular homes materialized. Road salt had bleached the asphalt and cracked the white shoulder lines. My heart began to wilt. This could have been Anyburg, New England. Where was the blood, the sweat, the glory? Where were the ghosts: John Kelley the Elder, who gave Heartbreak Hill its name, seven-time wreath-wearer Clarence DeMar, four-peater "Boston Billy" Rodgers, even disgraced subway rider Rosie Ruiz, names that had begun crowding my mind since I mailed in my registration check four months earlier.

I still wasn't exactly sure why I signed on for the Boston Marathon. I'd been running for six years, mostly solitary five-mile stretches on backcountry roads with sunrises so gorgeous they left bruises. The thought of pinning on a number, herding into a corral, and racing through crowded city streets seemed profane. But after moving from Middlebury, Vermont, to Cambridge, Massachusetts, the year before and settling so close to the sport's most celebrated course, I'd somehow been pulled into its magnetic force. I told people that at thirty-nine years old, a marathon was my version of the red convertible, a check mark on my bucket list. When pressed for a goal time, I'd answer to break four hours, other days to finish. I joked that my "secret" time was 3:51:38, one second faster than Pa's first marathon ("Nope, no daddy issues there," I smirked). But there was one goal I kept quiet. That hanging a finisher's medal on my wall would prove I was no longer the fiend I used to be, a 26.2-mile baton exchange where the present would finally take over from the past.

I never set out to be a runner, let alone a marathoner. Just as I never set out to be a drunk. And as I would learn over the next eighteen months, it wasn't just about taking the baton. It was about getting to know, and feel, the person handing it off—yourself—to take the stick without fumbling. And then learn to hand it off again, to let go.

When I kicked on a worn gym treadmill for the first time seven years earlier, I'd come upon a way to satisfy my urge to flee without actually running away, to exorcise my cowardice, to begin slowly drilling inward. After years of false starts and abrupt endings and burning shame, the accumulation of sweaty miles had started to make me feel capable, perhaps for the first time. Strapping on running shoes led to a reflection I didn't need to turn from. Without realizing it, I'd found another chance to become.

Would multiplying my normal run by five and performing among thousands of strangers on a very public city stage where I once behaved badly tattoo this effect? Would the storied history of the Boston Marathon, along with its pantheon of demigods and legends, all the worshipful hearts and personal stories and buzzing brains, feed into a single ink needle that would work on me for four-plus hours? Perhaps a deeper kind of becoming lay on the other side of pain, at the outer edge of my physical limits. An even clearer picture of who I was and why I'd acted the way I had.

It wasn't sobriety that led me to first lace up. It was vanity. The self-absorption and narcissism of addiction had followed me into the Big Dry. Without a drink Velcroed to my palm, I was finally able to kick a seventeen-year smoking habit. I filled the crater-size void with bacon pizzas and pints of Ben & Jerry's. The TV room always smelled like buttery microwaved popcorn, and the twisted foils of Ferrero chocolates littered the couch like golden roaches. Within two years, I'd larded on twenty-five pounds. When I saw a snapshot of myself on the beach the following summer, I was horrified. I looked like I'd swallowed a sack of Idaho potatoes, my misshapen belly pulling me toward the sand. The mass of flesh spread from my sagging man-breasts down over my suit, turning the waist-band over like a frown. I hardly looked like me. I could have been one of those poor bastards shown from the neck down on TV news reports about America's obesity problem. Was this really how life turned out? A handful of bad decisions and you're the fat, suburban dad in a tight pair of Dockers behind the wheel of a minivan, backing over your inner shirtless rock poet in low-waisted leather pants.

I joined a gym and found my way to a treadmill, tasting sweat for the first time in years. I liked the display panel be-cause it announced how many calories I'd burned, how fast I was going, how far, how steep. It broke me down into num-bers, giving my mind something to figure out. I saw a few pounds come off over the months. What I couldn't see was that those steps, that sweat, that pain, could scratch at some-thing beneath the fat, beyond the marrow, something more significant. But I hadn't yet logged enough miles to reach it. I was still getting used to a sensation I hadn't felt in years—for-ward motion.

But there was a lot of road that twisted between Those Days and These Days, and right now it was unfolding through an industrial stretch of Framingham, sliced with traffic lights and rail lines, the third town along the marathon course. All around, locals were grinding out their daily lives—men work-ing on utility poles, a tractor-trailer screeching its brakes, a cop with a car pulled over. I pictured last year's celebrity run-ners Will Ferrell and Lance Armstrong, the three of us run-ning neck and neck, cracking jokes, denying doping charges. But the image was snuffed out by the salty smell of French fries from a Wendy's across the street. Now I was just starv-ing. I made my way into Natick and scanned the horizon for

the CVS, a few hundred yards before mile 9. That's where I
planned to turn around. On my right, a rough breeze rippled
Fisk Pond.

I shook my arms out by my side, loosening my shoulders
and neck. Trees slid by and the wind pulled bits of tears from
the corners of my eyes. As I ground the pavement beneath my
treads, I could hear the swish of my nylon pants, the greed of
my hungry lungs casting its net in an ever-expanding rhythm.
Some runs, I got to breathing so deep and full, I was sure I
was pulling oxygen and light into some dusty, primordial corner, where perhaps gills once floated open and closed. That
sensation, coupled with the rush of my feet, filled me with
a sweet, swooning lightness and the feeling that I could run
anywhere, even up the side of the Prudential building.

At last, there it was: CVS, in all its red-lighted glory. Nine
miles had taken me an hour and twenty-eight minutes. By
this time, the elites, clocking sub-five-minute miles, would be
past mile 19, halfway up the Newton Hills. I turned into the
traffic wind. My heart sank a little at the thought of running
back through those same drab main drags. Then, without
warning, I started to get the "bacon feeling." I felt depleted,
then famished, my stomach scraping its walls for something,
anything. I'd hit empty. And all I could think about was bacon: crispy strips spitting on the griddle or draped across egg
yolks and buttered muffins. I could smell the burned gristle in
my sinuses, the back of my tongue aching. I could have mur-
dered someone for a Denny's "Grand Slam." I had to bite back
the anger rising at my body's sudden betrayal, its unexpected,
petty need. My calves and thighs felt filled with wet sand, my
arms coated with electric pinpricks, my skull a howling dust
bowl. Not unlike the feeling I once spent too many paychecks
on. Maybe I was too far ahead in my training, running eigh-

teen-milers at two months out. I'd been warned not to overdo it. I was still greedy, still wanted to be ahead of myself. But the more I fixated on my body, the fainter I became, my fuel belt an iron cummerbund, the distance between my foot strikes shrinking to baby steps. The bacon feeling had stopped my runs before. I found these meltdowns devastating, a reminder of the "incompletes" that once stamped my high school and college transcripts, the blank spaces in my resumé. To walk during a run, no matter the reason, felt like giving up, like watching prey grow smaller and vanish around a bend, perhaps forever. I danced between ten-ton buses and speeding motorcycles to cross Massachusetts Avenue at red lights. I hid my outrage when someone asked me for directions mid-run along Memorial Drive. Didn't they know? Unless you're on fire, I'm not stopping.

But what if I had to walk or, worse, drop out on race day? Failure to the fifth power. Would I just get on the T and head home? Fake-limp around the office the next day? Millions of eyes would witness my shame, my past tethered to me like a dogsled, branded a DNF (Did Not Finish). It wasn't lost on me that those three letters were an abbreviation of my last name. I ripped open my last two energy goo packets and throated down the gritty paste, the corners of my lips sticking together. After a quarter-mile, I felt some wisps of strength return, the sailboat righting itself. Maybe I'd pack some sausage links for race day.

Returning to Hopkinton was a lot harder than leaving. It was almost all uphill, plus I had to navigate blind corners that shot speeding pickups at me. I summited the last long rise and crossed back over the start line, slowing to a grateful shamble. I lived for that flush at the end of a run, a moment of near ecstasy when you finally allow yourself to walk, endorphins flooding your oxygenated brain, a beehive head, and the feeling that you've just done something worthy, that you've muscled through. That's what every run was—victory. I draped my belt of empty Gatorade containers over my shoulder like an animal pelt and floated to the pizza joint, and grabbed two vitamin waters, confidence and spirits lifted—18 miles down, just 8.2 more to master.

I spin the picture frame around on the kitchen table. The plastic is clouded over, licked countless times and gouged by razor blades, crushed granules spackling the nicks. Weather now permanently blows across the image. The photograph shows me, Mom, Pa, and Mandy in the Oval Office meeting with Ronald and Nancy Reagan. The only picture I have of us as a family. It isn't mounted, just slipped into a department-store frame with a cardboard back. The photo was taken a few days after Pa was released by the Soviets and deported. I'm sixteen, my eyes hidden behind long bangs, and I'm half smirking at the camera. My parents and older sister are dressed in crisp, formal clothes, blending seamlessly into the presidential scene. I'm wearing Chuck Taylors, slacks, and a paisley shirt. Mom and I had gotten into it that morning when I'd tugged on a pair of jeans. I'd fantasized about Reagan signing them next to the Led Zeppelin lyrics I'd scrawled above the knee. Mom threw up her hands and a family friend negotiated me into a pair of Brooks Brothers slacks and the paisley shirt.

Beneath the kaleidoscopic fabric, I'm wearing a T-shirt that reads "FREED NICK DANILOFF!!!" I had Magic-Markered the D and exclamation points after the word Free. The folks at U.S. News & World Report had shirts printed after my dad's first week in prison. In the holding area at Dulles airport, I'd grabbed a marker and made the edits. An hour later, fresh on American soil, Pa held the white shirt up and was photographed from vari-

ous angles, with me and Mandy, with Mom, the images appearing in countless newspapers and TV broadcasts. I'd scoured the captions and articles for a mention that "Caleb Daniloff, 16, had cleverly transformed the shirt." But nothing, no notice. By wearing the shirt to the White House, I was reclaiming my work. Once we moved to the Rose Garden for photos and questions from the press, I imagined ripping my button-down open like Clark Kent. When we shook hands, President Reagan seemed startled, as if he wasn't sure how I'd slipped past security. His large palm was dry and cool. On the shiny striped couch, I leaned forward to speak, but my voice came out on the same frequency as the room's ambient noise. No one heard. I was in a silent movie.

I run my moistened fingertip in the corners of the picture frame, dig a fingernail into one of the cuts in the plastic—nothing. There's no more denying the end has come. A car engine turns over, a thrown newspaper thuds against the stoop, the morning clearing its throat. The shades are drawn but I know what lies behind them: first light, the color of a corpse, as inevitable as a train, God's flashlight. I'm twenty years old, on leave from the University of Vermont. Freshman year hadn't gone as planned. Actually, it hadn't gone at all. I'm living with my parents in a small apartment in Somerville, working at a bookstore in Harvard Square. They're in Vermont for the long Thanksgiving weekend. I wipe the cold grease from my forehead. A bike I don't recognize leans against the stove. I have to be at work in a few hours. In the other apartments, decent folks are slipping from the warmth of their bed covers, brushing their teeth, starting coffeemakers. They terrify me. Even sober, I can never look them in the eye. I hold my breath. Are they hissing about me? Will they rush their kids past me on the stairs? I stare at the half-empty Budweiser with burned butts bobbing like dead bugs. The thought of stacking books and running a register, counting

out change to proper citizens with lunch dates and money mar-
ket accounts and 401Ks, sets my heart thrashing. I feel so alone I
wonder if I still have a reflection. I try to remember where Mom
keeps the sieve, but my brain only coughs and sputters. I pour the
beer into a coffee mug and pluck out the cigarettes. I'll be calling
in sick. Again.

It was early morning, three weeks before the gun. The
Boston Marathon is held every Patriots' Day, a state holiday
marking the Revolutionary battles of Concord and Lexing-
ton, though neither makes an appearance on the race route. I
was tackling the back part of the course. I'd started near Kenmore Square and the famous blinking Citgo sign and ran up
Brookline's bustling Beacon Street to the stony hilltop campus of Boston College in Newton, then down the Newton
Hills, and out to the Woodland T stop across from the country club. And back again. A twenty-two-miler. The farthest I'd
ever talked my legs into carrying me.

After fifteen miles, I found myself scanning my aches and
pains, my shirt a lobster bib of sweat. The scrape of my heavy
motion-control shoes and heave of my lungs began to tor-
ment me. I felt every one of my thirty-nine years. I was running without music, nothing but the thoughts in my head. I
might as well have been wearing one of those Orwellian head
cages filled with hungry rats. When a couple of Masters of the
Universe streaked by me in their body suits and pitiless sunglasses, I felt invisible, plodding in thick-soled running shoes,
an out-of-breath nurse jogging for a code blue. I made myself
feel better by barreling past an old man with crepe-paper skin.
He never knew what hit him.

At the Newton firehouse, I turned right, back onto Com-
monwealth Avenue, the wide boulevard split down the middle by a long stretch of green, with a vein of dirt track run-

ning through it. On either side were million-dollar homes and gleaming cars parked in circular driveways. This marks the start of the Newton Hills, a five-mile stretch with Heartbreak Hill at the top, poised for a swift kick to the groin. The hills themselves aren't monsters, but on race day they start cracking their knuckles at mile 17, the last thing your quads want to hear after a mostly downhill course.

In 1936, defending marathon champ John Kelley the Elder caught race leader Ellison "Tarzan" Brown on the last of the hills. As Kelley passed the Native American phenom, he gave Brown an "It's OK" pat on the shoulder. Brown took it as a taunt and popped his business into overdrive, pulling away for the win and, in the words of a *Boston Globe* reporter, "breaking Kelley's heart." That's where that Newton hill got its—falsely—intimidating name. People have come to expect Everest.

Even at 8:00 a.m., the hills were crawling with runners, like summer ants on a breadbox. Porta-Potties and merchandise tents had even been set up. This was my first glimpse of the field—all shapes and sizes and abilities, some in clusters, some solo. A pack of shirtless youths, thin and whippety, galloped past, feet barely touching the ground. A couple of them wore heart monitors, narrow black straps cinched around flat chests. Then an old man chugged by, tennis shirt tucked into his shorts, headband smushing his white hair up like a mushroom cap, and, of course, black socks up to his knees. I smiled. Up ahead, I stared at the clouds of sweat darkening a woman's sky-blue shirt, studying the heart-shaped marks like Rorschach symbols. In those heaving blots I saw everything and nothingness, a peaceful oblivion. Chugging forward, arms locked and swaying like train-engine wheels, it seemed I was still chasing the exhilaration of emptiness, that soothing pool of absence, except now I emerged renewed rather than with my lungs half filled with fluid and a negative bank balance.

But most of all, it was their faces. I loved looking at the softness in their eyes, the glaze on their foreheads like a pregnant glow. We waved at each other like old friends. *I see you. I see you, too.* Their hellos and snatches of conversation lifted me. I could hear their hard work over the car engines and barking dogs, their gasps and breaths, the pain lining the crisp air. We were all of us dreaming in unison, eyes wide open.

Making that first leap from gym treadmill to outdoor running some five years earlier, though, had been painful, not just physically. I felt like a tourist. In my mind, the outdoors was the terrain of lean, tanned bodies, specialized digital watches, butterfly shorts, and $120 shoes. But while on summer vacation on Long Beach Island, New Jersey, safely away from the eyes of Small Town, Vermont, where I'd moved after graduate school, I decided to give outdoor running a try. I'd built up to two miles on the treadmill without stopping. I pulled on my longboard trunks, strapped on a pair of backless running shoes that secured around the heel with a Velcro strap—sneaker sandals, really—that I'd picked up from the bargain bin for $19, and started off down the two-lane Beach Avenue. I half expected to feel the ground move and buckle, to hear the soft engine whir, ready to help move my 170-something pounds. The pavement was static and unforgiving. There were lots of other runners, bikers, Rollerbladers. The sun was blazing. There was no shade except in the shadows cast by the oversize SUVs parked in front of the bungalows. Every time a runner passed me, I felt they were saying, "You sure you belong here, kid?" I had no clue how far I should go, what an outdoor mile felt like, how fast I was going. I felt naked without my display panel, without my numbers. But I kept following the bike lane, the salty breeze on my lips. I even overtook someone. So what if she was etched in wrinkles and wore knee braces?

I didn't have a watch and lost track of time. The blocks all looked the same, pebble-filled yards, boogie boards drying against sides of houses, American flags flying, towels draped over porch railings, lobster-trap lawn ornaments. When the avenue merged into the causeway, I turned around, my shins starting to whine. When I got back to the house, I was shocked when I saw I'd been gone an hour and thirty-five minutes. I jumped in the car to clock my run. *Oh, my God.* I pulled over to make sure I was reading it right. *Seven miles!* When I got back and told my wife, Chris, her mouth dropped, a mix of awe and horror. Suddenly, I felt pain in my knee. Then beneath my heel. This wasn't good. Couldn't I do anything within reason? Now it was a waiting game. I could tell my body was already mounting a revolution, not waiting for sleep to attack. When I got up from the dinner table, my shins felt splintered. By morning, my lower back had locked up like a shuttered convenience store where someone had spray-painted FOOL. For the remaining five days of vacation, I was hobbled, grabbing onto furniture, mainlining Advil, and picking my way gingerly to the beach. I think I hated running. I just would never be good at it. I didn't have the body or the mind. It was flat and boring. Maybe I'd just get back in the pool, deal with the ear infections. *Hand me a chicken wing, will ya? No, the really big one.*

I join the knot of scruffy punks, businessmen, and a lone soccer mom in the basement, crowding around the group leader with our sheets as if straining for an autograph, all of us serving out our ninety-meetings-in-ninety-days sentences. This is num-

*ber 63 for me, all for a bottle of wine I'd stuffed down my pants
at a Shaw's supermarket. I'd been given Court Diversion, which
seals records for first-time offenders. I'm sure we're all thinking
the same thing:* Just sign the damn paper so I can get back to
doing what I was doing two hours earlier. *I wonder if this guy
hates us, feeling used for his pen. Or if we remind him of himself. I hold my breath and wipe at my nose, making sure no powder's caught on some booger that might have descended during
the Lord's Prayer. Please don't let him smell me. I need this signature. He looks me over as he scrawls his name on the list. I
smile weakly. I'm an affront to everyone in there; they're trying
to save their lives. I've been forced to parachute in and tolerate
the ninety minutes of testimony, a piece of Bubble Yum barely
covering my beer breath. I hate having to hold hands in a circle
at the end. I don't like the feel of a stranger's palm in mine but
don't have the courage to keep them in my pockets.* "Keep coming back," *whoever's next to me always says, giving my hand a
pump. As soon as I get my currency, I hightail it up the stairs, ten
feet tall, butt already out of the pack, as free as a summer-school
graduate, out onto Tremont Street. Evan and Sam are waiting
around the corner in the Vanagon. Evan hands me a pounder,
a fat line on a Tribe Called Quest disc, and a tightly rolled bill.
I'm touched at the greeting. I've been hanging out with them for
only a few months and still want them to like me. He pulls away
from the curb and the Boston night swallows us up, the streetlights lacquering the rain-slicked asphalt. I'm brimming, Q-Tip
softly rapping,* Bonita Applebum, I said you gotta put me on.

*Three hours later, I slam my fist against the windshield of the
Vanagon and watch it crack. I take a swing at Evan, too. I don't
know how I've arrived at this moment. I just know I'm angry.
I obey the feeling, follow the inexorable movement, helpless to do
anything but play the scene out. Evan tells me to get the fuck out.*

I fling the Vanagon door open and stagger bare-chested onto Mass
Ave. Or is it Beacon Street? Sam runs after me. I tell him off and
stumble back to Union Square, somehow losing a shoe along the
way. My parents are in Vermont for the weekend. Sam's girl-
friend, Nina, calls looking for him. I ramble semicoherently, tell
her about my fight with Evan. "He thinks he's so clever," I blurt.
Sam is convinced Nina and I have slept together, but we haven't.
Which suggests he doesn't trust her, which suggests I might have a
chance. I tell Nina I have a bunch of vodka. She doesn't bite, but
we chat for a while. I'm unaware that I'd picked up Nina's call
on the last ring; every slurring word is captured by the answering
machine, squeezed in between messages from the Nieman Foun-
dation Reunion Committee and my dad's dean at Northeastern.
My mealy-mouthed come-on eats up most of the tape. And waits
patiently to be played by my parents the next day.

Coming back from a morning run around the Charles,
I spied my wife, Chris, walking our dogs near our condo in
Cambridge's Central Square and chatting with a carpenter
working on an old home. He looked like he was trying to
make her laugh. And succeeding. Those construction guys are
all the same, I thought, and sped up. The guy stood about six
feet, wore a sweatshirt and Carhartt jacket, an apron of tools
tied around his waist. He looked rugged, with a ruddy face
and blast of white hair. Kinda old to be a player, I thought.

"Tom's run the Boston Marathon nine times," Chris told
me. "He's on the BAA."

"Really?" I said, looking at the ground sheepishly.

"Greatest event around," Tom said in a raspy Boston accent, then turned to Chris. "You need a number?"

From that day on, most of my weekday runs ended with a
debriefing with Tom Rooney and his Irish-born partner, Colin, who was usually up a ladder or hunched over the circular

saw. Tom asked how far I'd gone, about the road conditions, and capped things off with a story from one of his marathons.

"One year, I saw some friends at a bar in Cleveland Circle so I stopped in for a beer. I didn't care about my time. It's just so much fun out there. Another time, there was a guy running the race backward. Some guys run it barefoot. You'll have your own stories."

Tom gave me tips for avoiding bloody nipples and suggested I wipe Vaseline above my brows to channel the sweat from my eyes. "And everyone talks about 26 miles, but don't forget about that .2. It's the longest .2 miles you'll ever run."

Tom hooked Chris up with a pass to the VIP fete and finish area and said he'd get the MC to announce my name as I crossed the finish line. "Sure be nice to see an American win this year," he said.

I told him I'd see what I could do.

Rooney reminded me of another Irish laborer-runner I'd been reading about: 1917 Boston Marathon winner Bill Kennedy, a New York bricklayer. Depending on who's doing the telling, Bricklayer Bill hitchhiked or hopped a freight train to Beantown and slept on a pool table in a South End bar the night before. My kind of runner. Then with a Stars and Stripes bandanna on his head he won the freakin' thing in 2:28:37, the oldest winner to date at thirty-five years old, beating future Boston powerhouse Clarence DeMar. I was pretty sure Kennedy wasn't wearing a Gatorade fuel belt, with energy goo safety-pinned to a sweat-wicking shirt. Bill was beyond old-school—he laid the bricks that built the school. John Kelley called him a "rough character," but something about Boston must have smoothed his edges, at least for a couple of hours. He ran the race twenty-one more times.

"The lure of the Boston race is far greater than any in the

country and, to me, the world," he wrote to a friend. "They sometimes ask me why don't I give up marathoning. I tell 'em I'd rather give up bricklaying."

During the last few miles of his victorious 1917 run, sprinting down Beacon Street, an ear-splitting racket started to fill the air. It was his fellow "brickies," working atop a Coolidge Corner building, pounding their bricks together in applause. I loved that image. The night before the race, instead of painting my name on my shirt to tease shout-outs from the crowd, I wrote in large letters: BRICKIE.

Be the Brickie.

When I stumble to the kitchen, Mom is looking at me funny, eggs bubbling in the skillet. The Boston Sunday Globe *is scattered in sections on the table, toast crumbs on the magazine. There are paper towels and a bottle of 409 in the hallway. She asks if I'm OK. "I think I'm coming down with something," I say. "Yes, you look peaked," she says, turning off the burner. "You know, I woke up to the sound of you urinating in the hallway. You just looked at me. Do you remember?" Not a clue. I went to bed. It was black. I woke up. My heart's racing, fired like a pinball. "The Subaru reeks of booze and cigarettes," she says, her English accent like a thrush of birds chirping. "I think we're going to insist you go back to that psychiatrist in Waltham." I lower my head and nod OK. I feel like a bug pinned to a board. "Are you well enough for eggs?" This will all go away when I'm famous, I tell myself. I just need to write a few more poems. Or some cool short stories. Or a hit song.*

It was 6:30 a.m. and overcast, 49 degrees, a few hash marks below ideal running temps. This was it. Monday, April 20. Race day. A wall of school buses with tinted windows purred along the Boston Common across from the Cathedral Church of St. Paul, where I'd once attended AA meetings. Volunteers

in bright yellow jackets stood on planters and benches, buzzing megaphones at their mouths. Long, ragged veins of runners fed the buses. A giant fluorescent cloud of bodies floated
in front of the Porta-Potties. I scanned the sloping greens of
the Common filled with runners chatting, tying their shoes,
examining their bags. Beyond them, the benches were empty,
ghostly.

I took a seat on the bus among the rows of strangers all
dressed in sleek and shiny athletic gear, all on the same mis-
sion. Except most of them were Boston Qualifiers — 3:15 for
men my age, a 7:26 pace. That old feeling of not belonging
crept into my throat. A nine-minute miler, I got a number
through a charity. I wedged my bag between my feet and said
hello to my seatmate. The rows bubbled with murmuring
and laughter. Soldiers with rucksacks off to the front. It reminded me of the flotilla of buses that whisked me to Pioneer
camp when I was eleven. Pa had been assigned Moscow bureau chief for *U.S. News & World Report* in the early 1980s, a
five-year tour, and to turbocharge my getting-used-to-things,
a month after we arrived, I agreed to attend *Yolochka,* or Little Pine Tree, two hours outside of the city. My porous blad-
der had ruled out overnight camps back home, but, magically,
I'd been dry since we'd arrived in Russia. So early one summer
morning, I found myself bouncing along a potholed highway
in a caravan of loaf-shaped buses, heart thumping beneath an
itchy white Pioneer dress shirt, clutching a paperback copy of
Huckleberry Finn. My tongue could mold maybe seventy-five
Russian words. A hot wind off the road blew at my long hair.
The other boys babbled with each other and sported crew-
cuts and military-style field hats, necks splashed with crimson
scarves. Mine was bare, exposed. I stared at the road the same
way I was now watching the fog roll off the grass abutting

the Mass Pike. Same no-turning-back feeling. Except I knew more words and had a better sense of where I was going.

Some forty minutes later, we pulled off the interstate, state troopers manning the exits, ice cubes flashing. The Athletes' Village was set up behind the high school. Music blared through loudspeakers and clusters of runners had gathered under long white tents. A massive crowd was huddled in front of the blue Porta-Potties. I made my way into the woods. Among the trees, I came upon a handful of runners passing toilet paper and wipes, a band of Gypsies. Several women had their wispy nylon shorts down and stretched between their knees, audibly drilling the ground. Embarrassed, I turned my gaze the other way. Near the school steps a man arced a stream into the bushes, junk in full view. I relaxed and pulled down my waistband. It felt like a huge family, no leering, nothing we hadn't seen before. You left your shyness at home. This was the Boston Marathon, goddamn it.

I wandered back to the action behind the school and squatted against a brick wall. The PA was suddenly swallowed up by the two F-16s tearing open the cloudy sky. The elites were already lined up, birdlike, fingers on their watch buttons like pulse points. I dumped the contents of my bag on the ground, picking out my energy gels, pouring Gatorade into my plastic containers, then started stripping out of my sweatpants. "Caleb?" I looked up. It was a grinning Bill Coffey. "I'm just headed to my corral."

I'd met Bill, a ginger-haired police officer from outside Chicago, the night before at the "pasta event" at Boston City Hall. He was seventeen corrals ahead of me; he'd qualified with an intimidating 3:18. This was his first Boston, as well. Bill's mom, Mary Anne, had contacted me four years earlier after I wrote a commentary for Vermont Public Radio about a

Revolutionary War soldier named Zeeb Green, who marched under George Washington at Valley Forge and fought the British at Bunker Hill. Turned out Zeeb was Mary Anne's great-, great-, great-grandfather. She and Bill were genealogy buffs and had already trekked to Vermont to see Zeeb's grave. Mary Anne and I struck up a correspondence. A couple of months earlier, she'd let me know her son was running Boston, too.

Bill asked to take my picture but insisted I take off my sunglasses and hat, even brushed my hair down with his fingers. I think he knew Mary Anne would want to see my eyes. I still felt uncomfortable with good, decent people befriending me. I was afraid they'd detect something offensive and feel duped. "Smile," he said.

When Bruce calls me into his office and tells me I'm fired, I'm surprised. Not that I've never found myself in this position; this kind of entanglement with authority, this inability to meet my commitments, has become second nature. Since elementary school, I've been in a state of punishment—spankings, groundings, disciplinary probation, academic warnings, Court Diversion. No, I feel indignant because, even though it was a lie, I'd let my supervisor know I wasn't feeling well enough to come in. "You've missed nine shifts," Bruce says. "I'm sorry, Caleb." My name hangs in the air between us. I don't like it when people address me by name. It's both intimate and formal, like a smear of Icy Hot in my eye. Rob, the short musclebound security guard, who once told me he could kill a man with two fingers, follows me out of the bookstore and into the 7-Eleven around the corner. "What's next for you?" he asks, trying to sense whether I might come back and do something crazy. "I don't know, head cross-country," I say, trying to sound carefree and romantic. The next week, I get a severance check for two weeks' pay. I can't believe it.

Maybe Rob thinks I still need calming down. Does he see some-thing in me I don't? A person capable of revenge, violence? I don't know, but I like it and throw the perception over my shoulders like chain mail, another layer against the world.

I recognized the muted strains of the national anthem but never heard the gun, only cheers and whistles rolling back like waves. My heart was thumping, legs tingling in antici-pation. Runners jumped up and down to see up ahead, hop-ing to catch sight of a shift in the formation. I snugged my gloves and straightened the bib on my red Respite Center shirt. Number 23156. Then the crowd started budging, heads and shoulders breaking loose, shuffling forward like a gath-ering sneeze. Runners were tossing their sweatpants, bulky hats, and winter gloves. Volunteers just as quickly collected them for homeless missions and the Salvation Army. I care-fully draped my brown sweatshirt over a corral gate. It took fifteen minutes to reach the starting line, the white stripe now painted as bright as Elmer's Glue. In between the stomping feet, I looked for the BAA's blue unicorn logo, but all I saw was a blur of shoe logos, socks, and timing chips. Then I was carried across, tilted downward.

I couldn't see the road ahead of me, where it bent, where it dipped and flattened out. It hardly resembled my training run here a few months earlier. I couldn't see the houses, the shape of the woods. I just followed the stampede, running by faith. Despite the sea of heads bobbing as far as the eye could see, I could sense the downhill and held back, dropping to a ten-minute-mile pace. To break four hours, I'd need to average 9:07-minute miles. I'd make up for this later. Other runners, including a Santa, a Lucifer, and an English bobby, streaked by, carving out space on the soft shoulder. Halloween in run-ning shoes.

Along the way, bluegrass bands and barbecues were set up in restaurant parking lots, men holding beers in foam cozies stood at the roadside, coolers and tethered dogs at their feet. We were an excuse to drink at nine in the morning. I could almost taste those sour suds on my lips, a vaguely sweet ache at the back of my throat like a single grain of sugar dissolving on my glands. My pace quickened as I grew nostalgic. Pounding leftover beers or Jägermeister shots first thing after waking was magical, the sunlight streaming through the window, a lakeside reggae festival on tap, a chance to flirt or fight with exes, the day flung wide open. Isolated in time, the scenes made me feel free, full of possibility, adventure around every corner. It wasn't all bad, was it? I worried when I basked in the fondness of wasted memories. I'm sure there were fun times, but my brain tended to only dial up the battlefield — the fuck-ups, the spat words, the selfishness, the shame. Did affection for this part of my life make me vulnerable to relapse, disrespectful to those I'd harmed? Maybe just human, I told myself, fingering the sweat from behind my sunglasses.

Knots of sweaty charity runners galumphed past me, their damp T-shirts ironed with personal snapshots — a sister who'd died of breast cancer, a father failed by his kidneys, a child claimed by leukemia. The backs were frescoes: tributes and proclamations of love, RIPs, and "Never forgets." It was like reading the chapters of a communal book. I was awash in stories, my little boat lifted, oars moving in unison.

I was running for the Michael Carter Lisnow Respite Center in Hopkinton. It was the smallest of the Boston Marathon charities, and the money the seventy-five of us runners had raised made up a third of its annual budget. Michael Carter would have been twenty-three today. He'd been born four months premature, with cerebral palsy, blind and mute, weighing just over a pound. His father's wedding ring could slide up and down his arm. Doctors gave him a ten percent chance of blowing out the candles on his fourth birthday cake. "He understood," his mother, Sharon Lisnow, told me when I visited the center. "He just couldn't verbalize. He could laugh and he learned the sign for 'I love you.'"

"When I met Sharon, she looked like she had shoe polish under her eyes, like a football player," codirector Mary McQueeny told me as she walked me around the family-home-style building. "She didn't sleep for ten years. These mothers are living on the edge. You just never know what lies around the corner. Out of nowhere, Michael would aspirate, get pneumonia, and be in the hospital for three months. Or they walk in the bedroom and their child is dead."

It's a marathon with no finish line and no medals, and the toll can be devastating. Like 90 percent of marriages of parents of developmentally disabled kids, Sharon's marriage collapsed under the emotional and financial strain. Sharon and Mary began looking into opening a support center for families to relieve some of the burden and relentless worry.

On April 20, 1996, during his seventh stomach surgery, ten-year-old Michael died. It was the day of the hundredth anniversary of the Boston Marathon, an epic event with twice the normal number of runners, history on steroids. Beneath the celebratory din, a mourning was taking place, the deafening silence of hearts shattering. The next year, on the first anniversary of Michael's death, Sharon and Mary ran Boston in his memory.

The marathon has been keeping Michael alive ever since. For his part, Michael, dead now thirteen years, had allowed me and seventy-four others to reach the pinnacle, to run with the elites, the best the world had to offer, without qualifying.

I could feel my stride lengthen, a lightness infusing my legs. "Go, Brickie!" someone yelled, my first shout-out. I worked my pace down to 9:08.

In Natick, orderlies had wheeled the residents of a nursing home to the sidewalk, a row of them in thick recliners, swaddled in blankets, hairdos set, eyes twinkling, and clutching miniature American flags. I waved to each of them. We were fireworks exploding. Down the road, children and teenagers slapped hands with me and extended orange slices. Someone had built a wooden replica Fenway Park scoreboard and posted the latest score from the Sox game, taking place some fifteen miles away in Kenmore Square. A couple of dudes in red Speedos streaked by. I smiled.

Before I knew it, I was breathing down the neck of mile 14, a quarter-mile from the Scream Tunnel. You could already hear the rumble and drone of the "Wellesley Girls," wildly enthusiastic coeds from Wellesley College, five and six deep, straining over the barricades, holding pompoms and signs and shaking cowbells. Runners stopped for hugs and kisses. A French Revolutionary and a pair of giant red keg cups were getting a lot of love. I knew if I stopped, I'd never start up again. The cheering sound was almost tactile as I passed through, a tunnel carved of unrelenting, urgent human breath. By the time I reached the water station at mile 15, Deriba Merga of Ethiopia had broken the tape, with a time of 2:08:42. The race now belonged to the rest of us—the slow, the bruised, the broken, the unqualified.

I thought of Katie Lynch, all twenty-six inches of her. Katie had been born with a unique form of dwarfism, had survived countless high-risk surgeries to deal with her weak connective tissues, and was mostly wheelchair-dependent. "I want to run the Boston Marathon," she told race director Dave McGillivray when she showed up at his office. Not knowing quite how to respond, McGillivray said OK. "But I have a caveat," Katie continued. "My marathon isn't 26.2 miles. It's 26.2 feet."

Katie spent months training, undergoing water walking therapy at Children's Hospital six days a week and a program to strengthen her legs. She'd read all the marathon books, even carbo-loaded.

The morning of the race, McGillivray barricaded off a section of track from the start line 26.2 feet down Route 135. The media had gotten hold of the story and cameras were clicking away. Helicopters were circling, and VIPs and crowds of runners were gathering to see what the hubbub was about. McGillivray gave the signal and off she went, no higher than a step stool, clutching a miniature walker. McGillivray figured it'd take Katie about fifteen minutes, but when she staggered across the finish line in four minutes and into McGillivray's arms, the crowd erupted in cheering. There wasn't a dry eye in Hopkinton. "You have your Alberto Salazars, your Bill Rodgerses, your Uta Pippigs, all your marathon heroes, and not to take away from them, but this was the most emotional scene I ever experienced in all my years being involved in the Boston Marathon," McGillivray said.

McGillivray placed a laurel wreath on Katie's tiny head and draped a medal around her neck, congratulating her on being the race's first finisher. "She did it her way, but she ran her Boston Marathon."

McGillivray put her in the VIP section, then left to go about his race-day business. The last item on his agenda every year is to run the course himself, well after the last finisher has showered and the roads have reopened—in part because he adores the race (he ran it fourteen times before becoming director) and partly to make sure every town along the route has been cleaned up and left as found.

It was around 8 p.m. when he made that final turn onto Boylston Street. Up ahead, in the distance, he saw a wheelchair by the finish line in Copley Square. "At first, I didn't think anything of it, but as I got closer, I saw it was Katie. She was there waiting for me. After I broke the tape they put up for me, usually police tape or toilet paper, Katie came up to me. She put a laurel wreath that she made on my head and a medal around my neck. Then she looked up at me and said, 'Ha, I beat ya.' I will never forget that moment."

With each passing mile, the ever-increasing roar of the crowd, it became clear that the Boston Marathon went beyond finish lines, personal records, and the wall. It peeled open some kind of rushing, thumping, joyous, collective heart, a single organ shared by both runners and spectators. This course was paved with memory, prayer, generations of sweat, and healing—a Lourdes, of sorts. Here, once a year, mostly in the middle and back of the pack, the damaged gathered, put themselves through a trial, and emerged transformed. Maybe that could happen to me, too.

Boston had already reshaped me, at least on the outside. Over the past four months, I'd shed twenty-five pounds, dropped two pants sizes, and melted down my daddy belly. *Lighter equals faster, right?* People noticed my hollowed face, the saggy seat of my pants. After years of mainlining fried dumplings and mac and cheese, I scrutinized the hell out of nutrition labels, tallying calories and fat grams. Controlling these numbers somehow put the smell of a sub-four in my nostrils, drew me closer to completion, to perfection. I parted company with meat and dairy and learned to love tofu and broccoli. *Lighter means less pain.* I held my nose at people chomping on Costco hot dogs or rustling in McDonald's bags, once staples of my diet. I felt both affection and hatred for my scale. If I didn't like the numbers, I'd first curse the fat guy in the mirror, then try to soothe him. *Maybe the dirt and sweat streaking my calves are weighing you down.* I got off on my stomach grumbling, as committed to deprivation as I'd once been to indulgence. I learned to fall asleep hungry. The monk of Cambridgeport.

On the second of the Newton Hills, around mile 18, my dad popped out from the crowd and ran with me for a spell, in his leather jacket, paisley scarf, and highwater corduroys, his face grooved with lines. It was cute. I wondered if this was his way of making up for the missed baseball and football games. He asked how I was feeling, said maybe he'd run with me the next year. At seventy-five years old, and his last marathon almost thirty years in the rearview, I was pretty sure he didn't mean it. He rowed more than he ran these days, spending the early mornings plowing the waters of the Charles. Having been an oarsman at Harvard, he fell in love early with sculling. But as a career-focused journalist on the rise, putting in a shell on the Potomac or the Moscow River wasn't feasible, so he ran. Now a professor of journalism at Northeastern University, he was back on the water. On my morning runs around the Charles, I sometimes saw him slicing through the mist in his custom-made wooden single, carefully looking over his shoulder every few strokes. I'd whistle to him from the banks and he'd look around for a few moments, then spot me and nod his head. It felt apt, the mild manner of acknowledgment. But we both knew if he let go of an oar to wave he might capsize.

Before we moved to Russia in 1981, Pa used to run the Marine Corps Marathon in Washington, DC, tackling his first 26.2-miler at forty-one years old. By his last MCM, he'd qualified for Boston with a 3:19. But he never ran this course. The Moscow assignment got in the way. Got in the way of a lot of things. In Moscow, Pa and I became awkward with each other and didn't interact much except when I was being disciplined. I wondered if that's why I got in so much trouble.

When I thought of my dad, one of my most vivid images was his profile, hawk nose, thin dry lips, thick glasses and gray cardigan, heart clickety-clacking to the sound of his fingers punching the keys of the Telex machine, his lungs filled with the hot breath of his impatient editors back home. I hated that I'd followed him into journalism, but I was too lazy, or scared, to try anything else.

After a few hundred yards, he patted me on the back and peeled off. On my own again. I could taste the salt on my chin. I killed the last of my Gatorade and was already parched. I had fire hydrants for legs.

"You can do it, Brickie!"

By mile 23, I was dragging my carcass like the killer cyborg at the end of *The Terminator*. When I crested the rise leading into a thickly packed Kenmore Square, the red and blue Citgo sign throbbing over the fancy hotels and restaurants, my skin rippled in goosebumps beneath my clammy Under Armour top. I knew it was almost over. I was unnerved by the sudden welling of emotion, the crowd, the noise. My heart leaped, beating from a seemingly new place. The sidewalks were throbbing, seven and eight people deep, fans standing on coolers, kids on shoulders, whistling and cheering. Every year, the Red Sox game lets out from nearby Fenway Park so the multitude of fans in *B* ball caps and jerseys can transfer their energy to the passing runners. The battle was just about won; the troops were being welcomed home. Some limping, some walking, some crying.

"Move to your left, please." I heard a calm male voice behind me.

Startled, I turned to see a tall runner tethered to a female competitor, as if he were walking her. As they passed, I saw that the back of her shirt read BLIND RUNNER and his read GUIDE. I didn't understand what I was seeing. How was this possible? How could she not trip? Then they faded from view, dipping down beneath the Mass Ave overpass. A few minutes later I, too, sank below the roadway and popped out near Hereford Street, the next-to-last turn of the course.

As I passed Newbury Street, the crowds were frothing over the fences. The noise was deafening: the cheers, the shout-outs, the *Rocky* music blaring from a brownstone's second-floor speakers. "Brickie"s exploded in the air like Jiffy Pop. I turned onto Boylston Street, the triumphant final stretch, flags fluttering, people screaming and whistling from rooftops and balconies, banners and signs taped in office windows. I didn't even feel my body anymore, my senses overwhelmed, the pain and effort subdued by the crowd. I was floating.

Then I spotted Chris, all alone at the gate in front of the VIP bleachers, the elites having long since streaked past. Her short black hair spilled from her hat, bangs falling over her sunglasses. I stretched out my hand, then stopped, and we hugged. After a moment, I spun away, back onto the final twenty yards of the course, hurtling toward the bright yellow and blue line across from the Boston Public Library. I felt moved. By Chris, by what I was moments from completing, by the world. Just steps from the finish, I heard my name in the air: "Caleb Daniloff from Cambridge, Massachusetts."

Announced to the entire galaxy. Good old Tom Rooney. I hit the line and the soft timing mats. And just like that, it was over. I was walking. It was almost anticlimactic. I staggered into a foil cape and someone shoved a bottle of water into my hand.

As I moved through the waves of volunteers, I wasn't sure whether I'd truly been transformed or how much of my past I'd shed, if any. I knew I'd left part of me back there along the course. But I was also pretty sure something had opened up within those 26.2 miles. Doors, a field. I didn't want to lose sight. In the final hours, the pain had been excruciating, but I was ready to keep knocking, to feel for more hurt, for more edge. Maybe pain leads to clarity. My body tingled as if electric currents were making their final rounds within my crumpled limbs and wrung-out mind. Suddenly, a medal was draped around my neck. The curtain dropped, the lights came up. After all that training, I couldn't believe there was no more. Part of me wanted to do it all over again. I didn't want to wait another year. I knew Burlington, Vermont, where I'd stumbled through six years of college, staged a marathon the next month. Plenty of ghosts there, and hurt, too. So two days after conquering Heartbreak Hill, my thighs and shins still muttering, lungs grumpy, my heart regaining its shape, I signed up for the Vermont City Marathon.

4 hours, 20 minutes, 5 seconds
Average pace: 9:55-minute mile
13.1-mile split: 2:04:58
18,747th place out of 22,849 finishers

Burlington, Vermont

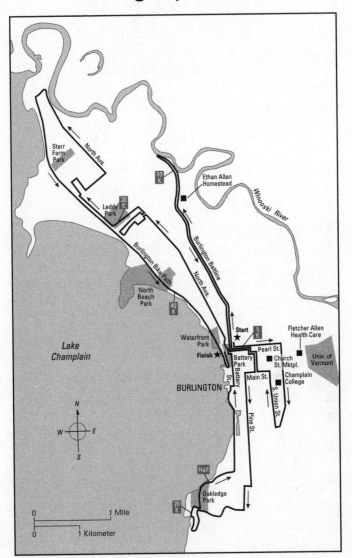

Starr Farm Park

North Ave

10 K

Ethan Allen Homestead

Winooski River

30 K

Leddy Park

Burlington Bike Path

Burlington Beltline

North Ave

North Beach Park

40 K

Lake Champlain

Waterfront Park

Finish

Start

5 K

Fletcher Allen Health Care

Pearl St.

Battery Park

Church St. Mktpl.

Univ. of Vermont

BURLINGTON

St.

Main St.

S. Union St.

Champlain College

Pine St.

N
W E
S

Half

20 K

Oakledge Park

0   1 Mile
0   1 Kilometer

2

21st KeyBank Vermont City Marathon and Marathon Relay

Sunday, May 24, 2009

THE RAIN WAS PULLING back and a mist hung over Battery Park. Beyond the low stone wall circling the green, the jagged peaks of New York's mighty Adirondacks rose up from Lake Champlain to meet Burlington's gray, brooding skies, muscles flexed, nostrils flaring, a showdown. I repinned my bib, moving it from my shirt to my shorts and back again. No place felt natural. For the next four or so hours, I'd be number 2915, one of three thousand registered runners.

This was my second marathon in five weeks. I was a bit surprised, and more than a little nervous, to be toeing another start line just a month after Boston. And not just because of my flat, asphalt-phobic feet and balsa shins. But once I'd stuck my BAA bib number on the fridge next to the Costco coupons and postcards, I'd felt a sudden loss. I'd heard that marathoners sometimes experience a postpartum-type depression after the race, but this was something more. I felt like I'd poked at part of me that hadn't stirred in a long while. The rigorous preparation, the physical strain, the deprivation and testing of boundaries, the public performance, all of it had combined to create a light-bulb flash in which I caught a glimpse of myself as I once was.

When I told my wife I was signing up for Burlington, my Boston gear still damp in the laundry basket, she'd given me a knowing look. "You can't just eat one burrito, you have to have five hundred," she said. But I was pretty sure I could wring another 26.2 miles from my training. Plus, the Vermont City Marathon literally bisected and dissected, crissed and crossed, a large part of an old sinning ground. There was hardly a square foot of town that hadn't been smeared with my slug trail. In many ways, the journey started here. Burlington was where most of the Indians were buried, the place I'd have to make peace with for all the others to rest.

As a college student in the late eighties and early nineties, I'd called Burlington home for six years and, as an ad agency copywriter, commuted here from Middlebury in the early 2000s, but I had no clue the marathon started at Battery Park, a former military camp built during the War of 1812. Now the grassy overlook mostly hosted concerts, ice skaters, food trucks, picnickers, and transients. The trigger on the inaugural Vermont City Marathon was squeezed my freshman year at the University of Vermont. But with an 8:00 a.m. gun time, the last runner was medaled and foiled before I staggered bleary-eyed to the bathroom to retch.

My parents nudged me toward UVM, keenly aware of my lack of aim, my taste for late nights, and my spotty high school career, which had ended in expulsion. They'd settled in southern Vermont after we left Moscow so Pa could write a book about his arrest by the KGB. He'd scored a fat advance, and they were building a house on a hill. We'd be Vermonters now. UVM accepted most in-state students and tuition was a hell of a lot cheaper for residents. So it was decided. I'd gotten into a few other schools, but it made little difference to me. I was ready for a new, looser phase of life. Plus, I knew a bunch of kids from high school who were going to UVM. It was an easy move. I craved easy.

The vibe at Groovy Uvey, as it was known, was electric, especially the first year. Students took over the president's office demanding diversity, protest shanties choked the main campus, a ragtag jam band named Phish was filling bar crowds with the spirit, and Burlington's mayor was a Brooklyn-bred socialist named Bernie Sanders. Long streets poured down from campus like asphalt tributaries into the black sparkle of Lake Champlain. It was a great place to get lost.

I landed in Patterson dormitory on Redstone Campus. I might as well have filled out a housing form where I checked off my passion for booze and asked if I could please be considered for housing where studying wouldn't get in the way. Double-headed beer funnels, Jell-O shots, beer balls, trashed rooms, released fire extinguishers. And that was Wednesday nights. I contributed by putting my hand through a glass hallway door reinforced with chicken wire, Jackson Pollock-ing three flights of stairs with my blood before campus paramedics tracked me down, dazed and woozy, in a women's bathroom two dorms over. In the classrooms, I didn't fare much better. I earned a C-minus in Russian, quite a feat considering I used to dream in that language. After fall semester, I dropped down to part-time so I could move off campus, closer to a new girl-friend and away from the eyes of the increasingly short-tempered RAs. An academic warning knocked on the door next, followed by a suggestion that I take some time off.

I jogged out past the police station on North Avenue to loosen up. The VCM route dipped in and out of downtown twice and covered the north and south sides of town, with several miles along the shores of Lake Champlain. A race em-cee and a one-legged elite triathlete pumped up the crowd from a low stage while prerace songs — "Born to Run," "Run-nin' Down a Dream," "Runnin' with the Devil" — blasted into the 59-degree air. I waited to hear if they'd play Pink Floyd's "Run Like Hell." It had rained the night before and show-ers were on tap for race morning. I fretted about blisters and chafing. I pasted Band-Aids across my nipples and rolled some anti-chafe gel on my inner thighs. A pole dancer warm-ing up backstage. But I didn't want to be the dude starting a forest fire with his chub or crossing the finish line with his man-boobs having burst into bloody tears.

With fifteen minutes left before the start, I squeezed into the crowd huddled in the middle of Battery Street. All around me, runners were chatting, joking, jumping in place, programming GPS wristwatches the size of laptops. One guy wore a T-shirt that read RUNDERLUST; another, BEER CHASER. I wondered if I had gone to school with any of them. I didn't look too closely just in case. I took a spot near the four-hour placard. Again, I was shooting for a sub-four. I'm not sure why I'd picked this marker, but I'd locked into it. Even though my 4:20 in Boston was plenty respectable — hell, just finishing 26.2 miles was — it felt on the wrong side of av-erage. Somehow, it wasn't enough, wasn't quite me. Someone from AA once said it takes five years to leave the bottle, five years to figure out who you are, and five years to become that person. I was almost at the ten-year mark. Would a 3:59:59 help speed me toward "that person"? Looking around, I didn't see as many charity T-shirts among the crowd and wondered whether those memorials in running shoes, that communal healing spirit so prevalent at Boston, might keep some of the doors closed and curtains drawn here. Burlington could end up being a long run and nothing more. I'd have to be OK with that, too. I decided to step back a bit, closer to the 4:15 runners, still afraid of the jinx.

Without warning, the crowd lurched forward. No gun, no horn. I didn't hear a national anthem, either. Good old Burly, not gonna salute just 'cause everyone else does. Or at least that's what I told myself as I began shuffling. I was glad to feel a bit of romance for the old burg. Within a couple of minutes, I was chugging with the pack, cheers bouncing off brick apartment houses, bodies coasting on clouds of unleashed energy and adrenaline. A young couple trotted past me, chatting about a work project as if they were still in bed with morning coffee. A group of college girls flowed by like water, singing snatches from a song I didn't recognize. The streets were wet and the grass shone.

As the herd wheeled onto Pearl Street, past the old courthouse and Bove's Italian restaurant, I moved to the outside of the clusters, weaving around two young mothers in Lycra capris jogging behind thick-wheeled baby strollers. Hard to think they were going to keep that up for 26.2 miles. They were chatting away, too. No one seemed concerned about what lay ahead. Except me.

At the corner of Pearl Street and Elmwood Avenue stood the post office, where one summer twenty years earlier I'd worked as a mailman, delivering letters and magazines in the old North End, met by impatient hands at the mailboxes every month for government checks. I once had the dog mace ripped from my bag ("I need it more than you," a stringy-haired woman in short shorts said as she walked off). I liked the noise and desperation of the old North End, the mostly French Canadian names taped to the dented black mailboxes, the intimacy of stepping into a musty living room with a blaring TV and bed-sheet curtains to have a package signed for. Whenever I delivered in the South End, sidestepping the manicured lawns and garden cranes and two-car garages, no one was ever home, not even a whisper.

Near the loading docks, a lone postal truck driver smoked a butt. One of the bay doors was open and a light shone. Maybe we'd known each other. But I couldn't make him out. A minute later, I was tilting toward my first apartment in Burlington, on the third floor of a white brick building, above the Other Place bar, a dark watering hole popular with bikers and postmen and college hippies.

It's 10:00 a.m. and the six-pack of Golden Anniversary is cold in the crook of my arm. Not too many people in the store and the clerk looks young. Just two customers ahead of me. I've been driving farther from campus to grab my morning-after beer. It'd be just my luck to run into a professor or one of my classmates. More people push through the doors; a few get in line behind me. What's the holdup? Some kind of discussion about lottery tickets. My heart picks up. I only have a credit card. That means a signature. One of my biggest fears: trying to write in front of other people—an impossibly delicate task with shaking, sweating hands, reeking of last night's booze, my motor skills shot to shit. More people join the line. Do I put the beer back and leave? I walked out of the last store because there were too many customers. "Next." Shit, I'm up. On stage, spotlit, all eyes on me for

sure. I slide the MasterCard across the counter. I feel sick, terri-
fied, like I'm choking. Sweat is pouring down my back. My chest
is jackhammering in anticipation, neck vibrating like an epilep-
tic's. The clerk looks alarmed as he runs the card. "Are you OK?"
It's getting worse. The machine spits out the slip, the performance
about to begin. I'm underwater, drowning at the bottom of a
well, the world far above me, unreachable. I'm going to break
apart if my heart doesn't slow down. Can people hear it? "I'm
fine." The clerk slides me the slip, staring at me. I lean forward.
My hand will not obey. I can hardly hold the pen; my whole arm
is convulsing. "Can you sign it for me?" I whisper, my tongue
like sandpaper. The clerk is perplexed. "Do you want me to call
somebody?" I put my other hand on top of my shaking hand and
scrawl two lines into the letter X under my name. "It's legal," I
mumble, grab my beer, and walk out, staring through the glass,
through the Chinese restaurant across the street. Before I even
start the car, I take a long pull off a Golden Anny can as if it's an
oxygen tank. And just breathe.

As cowbells clanged in the air and my arms punched
through the mist, waiting for my lungs to equalize and the
adrenaline to settle, I tried to pinpoint the moment, or sequence of moments, that led to my unconditional surrender
to alcohol. It felt as complex as an equation on an MIT black-
board. All I knew was that it happened here in Burlington. In
which house, which bar, on which street, I had no clue. From
the beginning, drinking had blissfully erased my shyness, my
social awkwardness, earned me friends. That float-brain feeling and the balls to crack wise felt wonderful. My problem
was I always wanted to feel better than wonderful—and then
just a bit better than that, my monkey finger furiously pressing the pleasure button. While drinking numbed my anxieties, it dulled the good parts, too—gratitude, joy, kindness.

Drinking shrank my other interests—music, reading, sports, even Russia—and one day while I wasn't looking, I took myself hostage, chloroformed and tossed in the dark, snuffing out all those things that when joined together gave me shape. They simply vanished. I'd issued no ransom note, made no demands.

This disappearing act started with relinquishing my memory. With my backward ball cap, frayed sweaters, paint-spattered jeans, and hiking boots, I blended in with my fellow Catamounts crawling home from the bars and puking in bushes (though I prided myself on an iron stomach; throwing up was strictly for amateurs). Like a lot of them, I didn't just get drunk, I got blotto, and chances were I ended up doing something stupid. I'd wake up alone in a pair of too-small women's sweatpants or on a strange couch with pennies and corn chips stuck to my cheeks. It was as if my mind and body had been hijacked, pushed into a bank with a machine gun like Patty Hearst. But I loved hearing about my exploits the next day—longneck in hand—as if I were my own twin, a perverse sensation that made me feel acutely alive, born again. "Dude, you remember last night?" became my "Once upon a time."

Meanwhile, my sense of accuracy became watered down until I began distrusting my own thoughts. Since my self-esteem was bound up in how knowledgeable I thought I was—and that meant having my facts straight—I was in a bit of a pickle. Pa, a career journalist with diplomas from Exeter, Harvard, and Oxford, had set the standard. He trucked exclusively in facts and information. He spoke like a newspaper article and was nothing if not stone-cold accurate. On occasion, someone would confront me on the street and I'd have little idea why. *Fuckin' kill you next time, dick!* Even though the evidence of my wrongdoing might have been circumstantial, I pleaded guilty every time. *I'm sorry, I didn't mean to.* I simply assumed I was in the wrong. Therefore, not a worthy person. Hanging my head became second nature. When I felt this kind of shitty about myself, the shots almost poured themselves. See how this works?

The clumps of pumping legs and swinging arms started to break apart. I picked up my pace, moved in and out of the empty pockets, taking to a few curbs, balance-beaming past folks. My breathing was unsettled. The mist had turned to sprinkles, coating my glasses. I tugged my cap over my brow and patted my chest to make sure the Band-Aids were in place. I felt a little ridiculous under my shirt, but then smiled. Yep, my outspoken nipples were still censored.

In my early days of running, I sometimes took to the Burlington Bike Path after work at the ad agency. Three miles tops. I hadn't wanted anyone to know I ran, let alone see me in my shiny shorts and futuristic running sandals, belly bouncing, my man-boobs a-jiggling. I was the new guy in the office, had never written ad copy before, and was back in Burlington for the first time sober. I still needed my boots and leather jacket, my armor in case I bumped into anyone I used to know. I was scared of anyone seeing me struggle so publicly, working to become something I wasn't. So I wriggled out of my jeans in my car, trying to keep the gearshift out of my ass and praying no one had parked nearby. I stepped out in my running shorts and shirt as if from a comic-book phone booth. Running Man. I always felt a little stupid.

Clutching a sweaty Discman, Eminem spitting at my eardrums, my hands flapping near my shoulders like dinosaur arms, I'd hit the wall at two and a half miles, clocking thirteen-minute miles. I didn't know what an orthotic was, had never heard of "pronation." All I knew was shin splints were as painful as they sounded. The day after, I was always semi-hobbled. Back-to-back runs were out of the question. Within months, I developed plantar fasciitis, a wicked gouge of pain on the sole of my foot, and had to sleep with a special white stocking that pulled my toes back like some kind of colonial torture apparatus. I was starting to think the run didn't care for the likes of me. It was monotonous and boring.

But still, I kept at it. Something in me wanted the pain, the drudgery. It fit with the efforts I was starting to make in my sobriety. Every limped step, every knee ache, was a switch to the back of my thighs as well as a call to awareness. Punishing myself became part of a healing process. The more I felt — physically and psychologically — the less things began to hurt. Though I wasn't quite aware of it at the time, inside my tiny universe some of the stars and planets were realigning; as slowly as teeth shifting in a jaw, perhaps, but there was movement. Little by little, I added an extra quarter-mile here, another ten minutes there, just past that tree, just around that bend. My stride started to gain fluency, my feet acting as metronome. The repetition became less monotony and more a rhythm of nothingness, like a Buddhist chant, my head humming with open space. On good runs, I disappeared, got truly lost.

By my second year of outdoor running I'd dropped my velociraptor arm swing and was logging five miles at a ten-and-a-half-minute pace. I packed away my scuffed leather boots and bought a pair of thick, wide-soled KEENs. My running and my sobriety had begun sharing a pace.

Just past Pearl Street Beverage, a couple of flannel-shirted bros had dragged frayed lawn chairs onto their second-floor roofs, sitting in low-brimmed ball caps, twelve-packs at their feet. It could have been 1992 again, guys I used to know stuck in time. Across the street stood my old dentist's office, where I had nineteen cavities filled my freshman year alone. I never flossed. Burlington was my Feral Period. I didn't know how to blow my nose, my sleeve cuffs covered in dried snot. I bit my fingernails and my oft-plugged left ear leaked a goopy wax, forcing me to walk on people's left side so I could hear. I didn't go near a fruit or vegetable and cooked copious amounts of burger, shuffling to the back of the classroom smelling like a short-order cook just off a double shift. I was surprised I didn't come down with scurvy. I was always broke, living off empties, bounced checks, credit cards I couldn't pay, and the meager proceeds from selling off my CD collection. Some days, I didn't speak to anyone except a phone-sex operator. But somehow, people expected me to be intellectual, civilized, worldly, educated, because I'd lived abroad, spoke another language, because I'd sat in the Oval Office with the president, because Pa was an international journalist. His reputation always preceded mine.

It's pitch black. What time is it? I'm confused. Someone is moving on top of me. My pants are down by my ankles. She feels heavy; I don't recognize the weight. I'm sweating despite what sounds like an air conditioner. I have no idea where I am. Nate's place? We've been going like this for I don't know how long. Can this be Andrea from my Stephen King class? Is it Casey? Am I cheating? I don't even know if I'm wearing a jimmy. I wish she would speak. My hands are on her hips, but I'm afraid to move them as if they're glued there. Long hair is now in my face, in my mouth. Her breath on my cheek. She knows me. I feel nauseous. I'm dropping, spinning, drowning.

The course turned left at Main Street and past Nectar's Lounge—a Mecca for Phish fans, gravy-fries lovers, and one

of my old favorite last-call haunts—followed by a tattoo par-
lor, coffee joints, and a vintage clothing shop. The spectators bubbling along the course reminded me of the eyes I always felt were on me when I lived here. Were they on me still?
Did they know my history—that I'd stolen empty kegs from
that fraternity, huffed whipped-cream containers at that convenience store, had dishes thrown at me in that apartment,
pleaded guilty to retail theft at that courthouse? The scenes
were moving fast, tripping over one another. I felt caught between needing to slow down and wanting to flee.

Before long, I was trucking along Route 127, Burlington's
Beltline, a four-mile hairpin loop, the first of several out-and-
backs on the course. My lungs had settled and my legs were
loose and strong. I was clocking 9:08-minute miles, about
where I needed to be. The leaders were already on their way
back—shirtless man-boys fluid as mercury, followed by lithe
antelopes with ponytails sweeping their tanned shoulders, di-
aling up sub-5:30-minute miles. Some of the runners on my
side cheered them on. The rain had picked up, soaking my
shirt and, to my dismay, washing away my makeshift nip-
ple protectors. I charged past a group of older women and
stepped into a puddle, drenching both feet. I pictured myself
all blisters and blood. Elvis dashed by in a white jumpsuit and
black pompadour, chased by a gangly teen in Chuck Taylors.

The loop ended with a long climb. I charged the incline,
drawing the moist air in as deep as possible, the chambers
of my heart feathering and squaring like oars in the water,
plunk, swish, pull. My philosophy on hills was to tackle them
with gusto, a chance to meet my limits and a swipe at former
fears and weaknesses. I was my own drill sergeant, pushing ca-
dets to collapse on sweltering summer fields. I obeyed the call,
embracing the burn in my lungs, coursing in my legs, spread-

ing through my brain. I pushed and battled. Who knew what lay on the other side? Perhaps disaster. Plunk, swish, pull.

I crested the rise along with some older runners and crossed Burlington's Manhattan Drive. I recognized the intersection. This was where Charles was killed several years ago. I scanned for the bloodstains I knew were long gone. We worked together only a short time. I was sober four years, afraid of my own shadow, when I took a copywriting job at Burch & Company, his College Street ad agency. At the time, Charles was battling liver cancer. His ex-wife and longtime creative partner had taken over day-to-day operations, but the craggy-faced seventy-two-year-old still went to client meetings. His spontaneous, wild-eyed ideas were comforting to the car dealership managers and chandelier company owners. He could write ads in the air. He was jazz with feathered hair.

Charles had already beat back pancreatic cancer and was staring down this latest set of fangs with a steely-eyed gaze. On the way to a client presentation, he insisted I feel the baseball-size welt on his side. In that brief touch, he brought it all home for me, life's struggles and triumphs.

Before moving to Vermont in the 1980s, his own self-imposed exile, Charles copy-chiefed at some of Manhattan's top shops. A high school dropout, he ate writing school grads like me for breakfast. Along the way, he'd raised up a serious booze and drug problem—of the wake-up-on-a-psych-ward variety—and burned through countless relationships. Decades after his last drink, Charles was still finding people to apologize to, would travel anywhere to say sorry. In all other areas of his life, he lived unapologetically, a freedom fighter. I thought about joining AA just to have him be my sponsor—and score some extra writing tips.

As we drove back from Montreal after recording a car

commercial one evening, Charles talked about offering himself as a servant to his victims. I was intrigued. I had just started reaching out to people in my past. I tried to imagine myself buffing Casey's boots. Or delivering Louisa's groceries. Or mowing Frances's lawn.

Charles spoke about some dark sins; one I wished he'd never told me. I struggled not to let that color my view of him. I never even told my wife. But it nagged at me, challenged my compassion. He committed it a long time ago, had made amends. And who was I to judge? I sometimes wondered if Charles had been testing my own capacity to forgive — myself, in particular.

It was an October evening when Charles was killed in a bike accident, struck by an SUV, all of his words gone just like that. I was stunned. I tried to find comfort in his refusing cancer the victory. I pictured him leaning his craggy face into the wind the way he leaned into life. The next morning, I laid a rose in the road, police tape still flapping off a telephone pole, blood on gray asphalt the last I saw of him. It had felt like such a mundane place to die. And it still did. But I was reminded, as I had been that day, that even in death, Charles had displayed his greatest gift: the teasing of poignancy from the blandest swatches of reality, his death scene a final lesson in animation.

I moved through the intersection. The rain neither picked up nor retreated and I was glad for it. Charles would have found giving in to such symbolism corny. I touched my fingers to my chest in silent prayer and strode on.

My girlfriend Casey's not around. Downstairs at the O.P.? On a beer run? I splash out more Cuervo shots for her younger sister, Courtney, and her friend Lisa, who are visiting. While Lisa's in the bathroom, I lean in to kiss Courtney. I don't know why. Because she looks like Casey? "Casey's my sister," she says, turning away, and walks to the bedroom. I cue up Jane's Addiction's "Summertime Rolls" for the third time and crank the volume. When Lisa returns, she and I smoke a couple of cigarettes and play quarters. She's gorgeous and tan. And terrible at drinking games. The next thing I know, we're in my roommate's bedroom and I'm climbing on top of Lisa, groping beneath her T-shirt in the dark. We kiss, she rubs my back. When I fumble with the top button of her shorts, she says we should stop. I stare at her face; even though it's dark, I can see the shine in her soft, catalog-pretty hair. "I don't really know you," she says. The sentence sobers me. Or rather its recited, timid tone. I don't hear the conviction in her voice, like it's something an older sister or aunt taught her to say at such a moment. And she got scared. She has no idea what she's doing. I roll off. Jesus Christ. She's fourteen years old.

At mile 9, the course wound back onto Church Street, spectators crowding the brick-lined pedestrian mall. My wife, Chris, stood at Leunig's Bistro, a French café I always felt was too fancy to drink at. She wore a cherry-red top so I could pick her out. I squeezed her hand as I passed. Chris was a runner, too. We always started out together but separated into our own paces. When we caught up with each other or passed coming back, we always clasped hands. It made a difference, feeling her in my palm. I was grateful she'd stuck with me, and when I thought about her during a run, I often came back in love. We'd met fifteen years earlier at a small newspaper in Middlebury, Vermont, where I'd taken a reporter's job after finally graduating UVM. She was a graphic designer and the mother of a four-year-old girl. She was in an unhappy marriage and would soon separate. Chris flashed easy smiles and had dark Cleopatra-style hair. She liked my leather jacket, and we shared a fascination with lurid crimes. She turned me on to Dylan. I turned her on to Bukowski. We sometimes grabbed drinks after work when Shea was with her dad. It wasn't for a year that we realized we'd gone to the same Massachusetts boarding school. Chris was the only person who knew my Before and After. I was still sometimes surprised she'd thrown in with me.

I let go of her hand and a few moments later hung a right onto Main Street and past the old Last Chance Saloon, a basement dive where a lot of folks found themselves at the end of the night, looking for a cheap drink, a pool game, a hookup, a fight. It had, appropriately I thought, been converted into theater space.

Farther on, a rockabilly band—stand-up bass, washboard, jug—wailed and thumped under the overhang of a ski and skate shop, also a former Phish haunt. Rain splashed off the windshields of parked cars, keeping time. I gained on an old man, his thin muscled legs pumping, tan skin loose around his elbows. I gave him a wave as I passed. He had clear blue eyes. He smiled and raised his hand. Gestures between runners reminded me of bikers waving at each other on the road. Harleys to crotch rockets, bearded tough guys to pudgy retired couples, it didn't matter. They were signals of brotherhood, of understanding, of shared passion. Knowing that I was moving among other like-minded people was comforting. I liked when runners said hello or thanked me for warning them of an ice patch or a loose dog. That moment of connection, however fleeting, was meaningful. There had been times when another runner or two had fallen in with me, and we chatted over a mile or so, about different marathon courses or prerun foods, then peeled off, back to our own paths. Was belonging as human a need, as essential as eating and fucking? Had my years of resistance to others, to my own friends and girlfriends, my punkass devotion to disconnect, dehumanized me? Can hearts stunted by selfishness and cynicism and anger ever regain their true shape?

A couple of miles later, the pack began rolling through a cul-de-sac neighborhood like a flash flood. The rain had let up and a strong sun began pulling apart the clouds. I could feel the back of my neck starting to bake. Families had set up tables in driveways and on lawns with drinks in wax-paper cups and bowls of fruit slices. One girl screamed and yelled and shook a homemade placard that read YOU CAN DO IT, UNCLE TIM!

I smiled, imagining one of those signs reading GO, ASS-HOLE! I don't know when the nickname got started, but for a good year, I answered to it—on the streets, at parties, in gro-cery stores. Badge of honor or dishonor, it was still a badge. I was foulmouthed and cutting and always juiced the stereo, played the same songs over and over (they seemed new to me at louder volumes). I remembered my good pal Nate first call-ing me "Asshole," as in "Hey, Asshole, put something different on, for chrissakes." Then he went casual: "Hey, Asshole, can you grab me an orange juice?" The name stuck and I didn't fight it. I'd always wanted a reputation more than I wanted to be me. Me sucked. Me was afraid to speak in class, terrified a cloud of stupid would roll out like morning fog. Me wasn't comfortable with "the guys." After a while, "Asshole" wore off and they started calling me "Kid Love." I liked that one better, fully aware of the irony, but content to let others define me.

"Damn, you see the tits on her?" a gravelly voice slurs, a rum-pled man gripping a wrinkled paper bag nudging his pal. It's Sat-urday morning the summer after graduation and I feel like a bag of broken glass. We're walking through Battery Park overlooking Lake Champlain. The man is talking about my girlfriend, Fran-

ces. I don't turn around. I don't say anything. I pretend I haven't heard, but his words attach to my swollen brain like barnacles. I steal a glance at Frances. Her eyes are fixed ahead. I don't dare look at her breasts. Surely, she heard. Is she expecting me to do something, say something, to defend her honor? I feel my face, my gut, burn with the pathetic truth. That I know I'll keep walking, my heart flopping like a caught fish on deck. Another piece of cowardice to add to the pile and the mental sweat of convincing myself it's OK: I didn't say anything because I'm not really in love; they were talking about someone else. It was a statement of biological fact after all, a compliment, even. But I know these are lame thoughts and I'm sure my yellow underbelly is shining like neon and Frances has discovered an ugly, essential truth. She's sleeping with a weasel. A boy. A useless loser on his way to growing a milk crate from his ass in front of 7-Eleven. How can she be with me? And what does that say about her? I light a cigarette with trembling fingers. Maybe I should break up with her. We're supposed to have dinner with her parents tonight. I'm gagged by the sudden appearance of their kind faces in my mind. But the farther we get from the park, and the closer to the change jar on my dresser, the easier I feel. Once I see 12:01, I'm hitting Pearl Street Beverage. Then, I'll be back. I'll show those guys. I'll show Frances what I'm made of.

At mile 15, Taiko drummers were thumping away at their skins at the bottom of Battery Street, stomping their feet, spurring the runners up the long hill. Another assault. Crowd noise wafted up from the waterfront like the smell from a barbecue. That's where the finish line was. But for the rest of us, it was still eleven miles away. An elite was already wandering about, lean limbs cabled with muscle, medallion around his neck, body glistening with achievement. His face looked serene. I wondered if he was the winner. It was only two hours, forty-some minutes in. He'd probably be massaged, showered, and burping chili by the time I hit the mats. I calibrated my pace with the deep, slave-driving rhythm and sailed to the top. Plunk, swish, pull.

By mile 20, I'd killed my Gatorade and was now relying on the drink stations. I didn't like not being in control. We dipped in and out of a few more neighborhoods before being shot back onto a main drag. North Avenue? I was disoriented, not sure where I was in relation to downtown. Despite my years in Burlington, I'd never seen these roads before. Where the fuck was the lake? My spit was so thick I could chew it and my legs felt like frozen sides of beef pummeled by a boxer. I was in new territory—and not just physically. A desert of numbness spread all around that I needed to push my way through, the sun toasting my shoulders and neck. Salt lines squiggled across my chest like an EKG readout. I kept leaning forward, following the motion of my body. Dumb momentum was my only strategy now, something I distinctly lacked during my college years. A young dude with a Mohawk gained on me and held out some apple slices as he drew even. I felt bad waving him off, passing up a chance for a brief connection. I still felt that impulse to resist, that feeling that even small talk was too intimate. But this time, I recognized it. And I was aware that I was aware. Panting and coated in sweat and parched throat screaming for a cool mountain brook, I was hyper-awake, all my senses on fire.

Despite the thinning cheers, I felt a surge of energy when the course turned off North Avenue and onto the Burlington Bike Path, past Starr Farm Beach and Leddy Park. This was the last throes of the race, the final four miles. I knew this stretch well and all the contours that would lead to the finish. Walking or running or just hanging out, the lakefront had

been one of my favorite spots in Vermont, the blue-black water dotted with boats and kayaks and lumbering ferries, the
long wall of toothy Adirondack peaks seemingly keeping the
rest of the country at bay. I scuffed down the squiggle of dirt
that snaked beside the asphalt, beneath a shady tunnel of leafy
tree branches and forested hills spilling down to the asphalt.
I'd been averaging 9:15-minute miles and was tiring fast. Re-
lay runners on fresh legs zigged and zagged, darting into open
pockets, tearing by us exhausted marathoners willing our way
forward, interrupting our rhythm, testing what little mettle
we had left.

A couple of bare-chested men in ball caps sucked on cigarettes by their fishing poles at the water. When I'd been a
student here, the shoreline used to be wilder, lots of scrub,
homeless camps, and a rusting grain tower. It had since been
developed, a boardwalk laid down with swinging benches, an
aquarium opened, greens cleared and landscaped, luxury con-
dos built with spectacular views. "There's no place to fuck
down there anymore," Charles had said.

Across from the dog park, I recognized the old brown
Dumpster sitting against the stone wall below the defunct
train tracks. Once during a run, I saw a disheveled man lean
his bike against the steel container, pull back one section
of lid, and peer in. Looking for bottles? A meal? A tarp? As
I shuffled closer, he moved to the right and raised the second section of lid. Scrawled on its underside in giant white
letters: JESUS WASTED HIS TIME ON US. The man looked
into the black for a moment, dropped the top, and rode on
empty-handed.

*It's late Monday afternoon and Big D and Frances are driv-
ing me to Fletcher Allen Hospital. I've told Frances I want to*

stop drinking. There'd been smashed crockery this weekend, complaints from the neighbors, a misplaced car. A mix of worry and relief darkens Frances's brown eyes. "You're doing the right thing," she whispers, and puts her hand on my thigh. D pulls into a spot and holds out the rest of his Camel Lights. I squirm as I take the pack, a transaction that suggests I might be here a while. I hate meanings. Frances checks me in at the intake station while I wait in a hard plastic chair, holding my chin in my hands, staring at the carpet pattern until it blurs. I'm ready, I'm ready, I tell myself like a ballplayer getting pumped in the locker room.

A nurse comes out and leads me to an exam room, takes my temperature and blood pressure, asks how often I drink, when my last drink was. I try not to lie and tell her I'm interested in a program. She says a doctor will be in to see me shortly and talk about next steps. I wait. The fluorescent lights become brighter. I open drawers to pass the time, turn over some packets of gauze and wooden tongue depressors. I can almost feel the last of the booze molecules evaporating. I get cold. I wait another ten minutes, staring at the pain chart with a rainbow of cartoon faces and numbers beneath them. I don't feel anything. Just alone. Very much alone. Where the hell is the doctor? The quiet becomes deafening, the lights sharper, blinding. I'm choking on my own presence.

Suddenly, the thought of giving up drinking feels horrifying. I like having an affliction. It makes me feel special. Who would I be without the stagger and slur and filthy mouth and disappointments and the heat of a disapproving world upon me? I was in love with my own anger and disaffection. I'd grown accustomed to the drama, the denials, the arguments, the makeup sex. Beyond all that was an abyss, at the center of which was total heartbreak.

An hour ago, when I was still buzzed from a morning forty, I could face the facts, say the words: I want help. I can't even appreciate the irony that booze gave me the courage to face myself out loud. Fuck it. I hop off the table. I open the door, speed walk down the linoleum hallway, past a surprised D and Frances in the waiting room, and out the door. Things will line back up tomorrow, certainly by Wednesday or Thursday. Maybe I just need some sleep and something to eat and to watch a little TV. I walk home alone in the dusky night, D's cigarettes floating somewhere in my coat pocket.

After a quarter-mile, the path entered the main section of the waterfront green where Nate got married in 2005. I was several years sober and former friends and drinking pals were on hand. Everyone was drunk but me. This was the first wedding I'd gone to since I'd slinked out on Evan's a couple of years earlier on the Jersey Shore. I'd been just nine months sober. I'd stayed for the vows, but it was unsettling to watch someone into whose mouth I used to blow coke smoke, intimate as a kiss, slip into adulthood so effortlessly. Was I jealous? Or resentful that Evan hadn't had to give up anything like I had, no voids in his life. In fact, he'd only gained. And somehow I knew he always would. Once the ceremony ended and people I used to know started talking to me, I froze, got scared, and made a beeline for my car. Sober, I had no idea who I was and I didn't want other people discovering it for me.

Despite our history, I wasn't in Nate's wedding party, either. In college, Nate and I had promised each other all sorts of things—to be each other's best man and godfather to each other's kids, to share forties on a porch in our rocking chairs. He was one of my oldest friends, an international kid, too, South African by birth and the adopted black son of a white

UN man stationed in Geneva. I envied his social ease. But after graduation, we saw less of each other. He made new friends, clean, success-plated guys I had trouble relating to. And in the end, I eloped and he chose a different best man. Under the reception tent as he and his brothers sang "Brown Eyed Girl" with the band, I knew I was watching us split into something else, and so I raised my glass of seltzer to the hand of fate—content now to be best friend emeritus—and led Chris out to the dance floor.

Just that .2 miles left now. I was on track to break 4:05, not quite my goal time, but fifteen whole minutes faster than Boston. And at this point, I'd take it. I felt spent, legs weighted down like a mob snitch. The crowds were thick and noisy, leaning over the orange snow fencing. I slapped a few palms, looking to dig a little deeper, to put off surrender a moment longer. It felt good, those strangers' hands, my fingertips morphing into mini booster rockets bringing me home. My hands sliced the air, the chafe under my left arm stinging as I made the final turn. Finishers' names were being called out over the PA. I was briefly disappointed when I didn't hear mine. I was unclaimed, no longer a native son, as if Burlington had dispensed with me. Or had I been released? I angled onto the straightaway and within moments, I was across the line. The action was all behind me now. Every thought, every molecule of feeling, was shimmering back there somewhere on the course, slipped from my stilled fingers, but I knew some had gotten under my nails. Volunteers urged me forward to avoid bottlenecking.

I was instructed to put my foot on one of the low stools and a young blond in a staff T-shirt sitting on a bucket undid my laces, congratulating me as she snaked the timing chip off my shoe and tossed it in a box like a beetle. Then laced me back up. I felt like a leper having my feet washed, the race draining from my limbs, from behind my eyes. I thanked her and grabbed a water, then headed into the concert-thick crowd, a little dazed, stiffening, my new version of the hangover starting to settle. Music blared from speakers and the smoky smell of grilled meat wafted through the air. Plenty of hands clutched plenty of Long Trail and Ben & Jerry's. The crowd glowed with runners in silver foils, finishers' hardware glinting around their necks. We'd all gone to heaven, a runner's paradise.

I patted my chest to take a closer look at the medal; nothing there. I'd spaced it. I staggered back into the pen, pushing against the current of runners pouring out of the chutes, and found a volunteer with dozens of medallions hanging from each arm like fringed sleeves. I didn't say a word, just nodded. For a moment, I worried he thought I was trying to pull a fast one. But he didn't blink, slipping a medal off his arm and parting the ribbon with his fingers to accommodate my sweaty head. I bowed, leaned through, and took the weight on my neck. I felt taller, evolved somehow. Boston wasn't a one-off. I felt I could now truly call myself a marathoner. But I was standing in another kind of light, too. After so many checks in Burlington's "no" column, the marks spilling into the margins, I'd just lodged a big one on the "yes" side. It might not even the score, but it brought my side of the seesaw a little farther off the ground. I headed for the train tracks, a shortcut to the parking lot, to find Chris. The thick gravel and rail ties shifting beneath my running shoes threatened to petrify my cramping calves. I saw Chris heading my way, so I stopped and leaned against the iron fence until she reached me. Despite the babbling crowds, the windswept music, and the tangle of cars trying to exit the lot, I felt perfectly still. Just breathing in the cool lake air, and breathing it out again.

4 hours, 4 minutes, 37 seconds
Average pace: 9:20-minute mile
Average speed: 6.4 miles per hour
1,076th place out of 2,335 finishers

Moscow

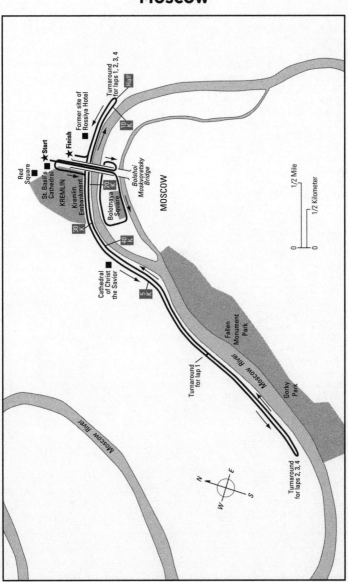

3

29th Asics Moscow International Peace Marathon and 10K

Sunday, September 13, 2009

KOLYA PHONED KOSH and said I was back in town, that he should meet us for dinner, "just like old times, the three musketeers together again." Kosh wasn't feeling well and said let's talk in a few days. Kolya slipped his cell back into his pocket and glanced out the car window at Moscow's standstill traffic. "Kosh's trying to stay sober," he said.

It had been fifteen years since I'd last been in Moscow, a summer intern at the *Moscow Times* after graduating from college. When I'd lived here with my family back in the eighties, Kosh, Kolya, and I had been inseparable, banging out music in apartment cellars, tipping back bottles in stairwells, grilling *shashlik* in the woods. Kolya's wife, Olya, shook her head. "Kosh's a coward." Her thick braid brushed the leather headrest as she steered the car into an opening. "When he's not drinking, he hardly talks. But when he's drunk, he's Mr. Party."

I shifted in my seat. She could have been talking about me.

I'd been surprised to learn that Moscow held a marathon, let alone for the past twenty-nine years. But I knew I had to run it. When we'd lived here, I'd rarely seen anyone running on the streets. Pa used to jog out to the Lenin Hills behind our apartment complex, and he was often met with stares. The Soviet Union prided itself on athletic prowess, but the emphasis was on team sports, the collective rather than the individual. But it was precisely that quality that drew me to running, its personal nature. It is the most democratic of sports, open to anyone, any age—just slap on some shoes (and even those aren't necessary) and head out the door, in any direction, for as long as you want, your mind unloosed from your brain, free to think and feel anything. The only person you have to measure up to is yourself. It is the essence of freedom, your destiny in your swinging hands. In Cold War Russia, that was practically a political act.

Today, the race was called the Moscow International Peace Marathon and started and ended just below Red Square. The course unfolded along the embankment road above the Moscow River, passing the Kremlin battlement walls, the Cathedral of Christ the Savior, Gorky Park, and the enormous statue of Peter the Great, the seventeenth-century tsar who founded St. Petersburg and the Russian navy. The route was adjusted in the 1990s when Fred Lebow, father of the New York City Marathon, himself born in nearby Romania, got involved to help grow the race, which had been limping along for the previous decade. The gun first went off during the Moscow summer Olympics in 1980, nine months before our arrival. Nine months before my life took a very different course, a sea change at eleven years old. While I developed a taste for Stolichnaya and blackout drinking here, this wasn't where I became a drunk. That was youthful fun rather than mind-numbing escapism. But Moscow was where I became the person who became the drunk, where a clock stopped ticking and a hole had opened up. I was back now, in a pair of Brooks Beasts and City Sports shorts, to try and close it.

"Never heard of the Moscow marathon," Kolya said. "What's the point of running forty-two kilometers?"

I thought for a second, but my Russian wasn't coming back as strongly as it used to, the right words eluding me.

"It's like an exam," I said, with a weak smile.

I once spoke perfect Russian, dreamed in Russian, but the language had been frozen in place, stilled when I was sixteen years old, when we left for good. Now, I could express myself only at that stunted level, a nervous teenage punk flailing around inside a cleaned-up thirty-nine-year-old body. I hated that feeling of not being able to communicate, a stroke victim. I felt self-conscious. And Kolya's video camera wasn't helping. From the moment I strode down the airport concourse, Kolya was zooming in on the barely contained anxiety writhing beneath my face, in my heart. Me, the unwilling star of a documentary: *Alone, Sober, and Out of Sorts: Caleb's Return to Moscow.* It felt like it might be a long seven days.

This was my first marathon without Chris, who had just taken a graphic design job at MIT. She'd been part of all my other races, my support crew. It made a big difference, knowing she was watching, showing her the medal, talking with her about the experience on the car ride home. I still liked to impress her. She'd stayed behind with Shea, who was starting her senior year of high school and was still struggling with the transition from her small, all-white high school in Vermont to the large urban public high school in Cambridge. Meeting my old friends and haunts would have to wait. I was nervous. Not drinking was always easier when Chris was around. A lot of things were. I'd been back to Moscow a few times, but never sober. When I'd quit drinking ten years earlier, I'd thrown a blanket over Russia, too.

The Black Eyed Peas' "My Humps" thumped from the radio speakers. A couple of Ducatis zipped around a clump of Volvo taxis and a Mercedes SUV. Designer storefronts crowded the busy sidewalks. All but a few Soviet-style shops with names like Milk or Bread or House of Shoes were long gone. Last time I'd rolled down these boulevards, there were far more open car lengths, and I was twenty-four, seeing double, clouds of tobacco smoke for breath. But that was a different Moscow, the flash and glitter and kiosk porn just starting to bloom, a cautious exuberance. This was some kind of chemistry experiment as if the city had swallowed a batch of Kwik-Modern concentrate, the streets wallpapered over with smoked glass and chrome and furs and tits-and-ass and gleaming hood ornaments.

From the car window, I watched a man in a business suit blow his nose onto the sidewalk without breaking stride, just put his thumb to his nostril, tilt his head, and let fly. A Moscow Hanky. A common sight in the Soviet days. Among runners, it was called a snot rocket. The lack of Kleenex was no big deal; who needed tissue when you had breath? Snot rockets were one of the few things that still reminded me of my Moscow years.

Almost thirty years earlier, in the spring of 1981, Pa, Mom, and I moved to this city. I was eleven, pulled out of Phoebe Hearst Elementary School, in Northwest Washington, DC, where I was a sixth-grader. My seventeen-year-old sister, Mandy, didn't move with us. She flew off to Northwestern University outside Chicago, assigned a guardian. Just like that, she became an orphan and I an only child. My best friend, Ethan, our baseball team and candlelit D&D sessions, Sunday Redskins' games, my duckpin bowling league, Daredevil comics, hot bags of Roy Rogers fries, *Cheap Trick at Budokan*, were all cleaved away as smoothly as if by a guillotine blade.

Moscow's boulevards were festooned with red banners and images of Socialist leader Vladimir Lenin and Party Secretary Leonid Brezhnev. No pizza, no hamburgers, no Coke. My last name, which had given American mouths trouble back home, now sang sweetly off Soviet tongues—*Daneelov*. The way it had been uttered for generations, before my great-grandfather, a military adviser to Tsar Nicholas II, saw the bloody writing on the wall in 1917 and sent his kids out of the country. I even received a patronymic—Nikolayevich—like all Russians. Pa was now part of my name. I thought that was kind of cool, like a title. I was now Kalyeb Nikolayevich Danilov. My first name was pronounced differently, a short, guttural *a* designed to make my name sound more Russian, like Kolya or Oleg. The syllables rose from the back of my throat in a deeper, manlier tone. A new boy poured into a bottle with a label everyone could read. Even my sniffing tic and bed-wetting had hit the butcher shop floor.

Yolochka *is shocking in its dullness. Mom and Pa think Pioneer camp is a good idea since they want me to attend Soviet school in the fall. I don't know much Russian, but another correspondent's kid, Colin, son of the* Chicago Tribune *man, speaks fluently and is supposed to be here with me, but he pulled out the night before. There are no video games or jukeboxes. The cafeteria offers up oily soups, wet chunks of meat, black bread, and a mysterious thick fruit compote. The pool is covered in algae. Activities involve Ping-Pong, volleyball, and uborka uchastka, cleaning up the assigned area—in our case, a section of driveway and yard around the three-story dormitory. We gather for lineika every morning to salute the Soviet flag and sing songs. I'm never told I don't have to raise my hand; I just don't. The only one of hundreds with both hands at his side. The only one wearing Levi's cutoffs, bare-chested beneath a velour hoodie, a head full of shaggy hair.*

During afternoon nap time, I hear some whispering and a blond crewcut boy pops out of bed and stands against the pale green dormitory wall. A second kid with a red brush-top throws off his covers and positions himself in front. Sunlight is pouring through the windows. The blond begins breathing deeply, maybe ten times. Others tiptoe over. I sit up. The boy then holds his breath. The redhead presses both hands on his chest and pushes firmly. Within seconds, the boy slumps to the ground. I stare wide-eyed. The blond boy looks asleep, peaceful. About a half minute passes before he comes to, looking around groggily. "Vot eto da, rebyata." I understand only the words yes and guys. The boy stands, dusts himself off, and walks to the end of the line that has formed.

I push my blankets off. I unzip my hoodie. There's no need to speak. They look at me funny at first, but then give me a what-the-hell look. The boy flashes ten fingers at me. I breathe as deeply as I can, my head growing soft as shaving foam. Then I gulp in one last swig of air and hold it just like the blond boy. A pair of small hands clamp my solar plexus and press me tight into the wall. Pinpricks, my brain shoved up through a keyhole, then everything breaks apart, darkness. When I wake, I see bare legs and stocking feet. I have no idea where I am at first or how much time has passed. I'm sure I've had a full dream. I stand up and

we all smile at each other. I get back in line. We pass out until
dinnertime.

A couple of days later, I called Kosh. His wife, Yulya, answered. Kolya had warned me Yulya could be unpleasant with Kosh's friends, especially if she suspected they had bender potential. A fifteen-year reunion with an old metal-head pal, I suspected, would have enormous potential in her eyes. "Who's asking?" Yulya said. "This is Kalyeb, a friend from America." She put the phone down. I heard a muffled *"Droog s Ameriki."* Yulya returned to the phone and said in a flat voice, "Lyosha's in the shower. He'll be free in ten minutes. You can call back."

Kosh's real name was Lyosha. But he's always been known as Kosh, a play on the Russian word for "cat." He had a springy way of moving, whether walking the streets or running his fingers up and down a guitar fretboard. And the nine-lives thing wasn't too far off. He'd survived car crashes and encounters with the *Lyubertsi,* anti-Western bodybuilders who harassed Moscow's longhairs, and, thanks to his musical chops, had dodged the rampant hazing during his conscription. Kolya even told me Kosh had been exposed to radiation while serving in Kamchatka.

Fifteen minutes later, I was met with a busy signal. I waited ten more minutes and hit redial. The phone finally rang, but no one picked up. I was getting pissed. He was clearly avoiding me. Then I pictured Kosh sitting on a couch, in his towel, fretting about whether to grab the receiver and how that simple, split-second action might ruin his resolve, change his life. I could relate. People fall off the wagon all the time, whether after eleven years or five months. He had a young daughter now. There was also shame in having to insult me. It wasn't the Russian way. But I was OK that he'd rather preserve his sobriety, his relations with his family, than see an old friend.

I'd given up friends, too. I tried calling one last time. Kosh picked up. He sounded happy to hear from me. "Come on over," he said.

I hopped on a trolley bus. The passengers seemed out of place, stuck in time. They were the first group I'd come across who looked even remotely like the Muscovites of my youth. Hard faces, net bags stuffed with vegetables, dated floral dresses, gold teeth. Here was the missing link between the former USSR and Russia, 2009: public transportation. They were trapped in Soviet amber. People who didn't have cars, didn't have dachas, who'd been left behind. Olya said she hadn't taken the subway in years. "I'd rather sit in the comfort of my car listening to the radio in a *probka* [traffic jam] than pack myself into a hot, smelly train car."

"Vashink taim," Directreesa says to me, holding a clear orange clipboard.

I'm standing among a crowd of boys in front of a long concrete building, many of them older. The sun is strong. I'm not sure what's happening. Beneath my velour pullover, my heart is thudding. It sounds like she said, "Washington." Why would she know where I used to live and what does that have to do with anything? A door opens and the stream of bodies snakes in. I hear a low pounding like muffled rain, echoes of voices being swallowed by the dark of the building's interior. Then it hits me. "Washing time!" For a moment, I'm pleased I've figured it out, the stiffness leaving my body. I follow the crowd, wondering if they'll have stalls or curtains. I've never used a public shower. One of the male counselors is handing towels to boys. I hear voices and water smacking against concrete, the sound not separated by anything, just a mass of noise. It smells sour. I don't like the feel of this suddenly. A group of younger Pioneers, maybe seven or eight, straggles in after me. The Directreesa closes the door and shoos the little boys, now shirtless, past me. No more sunlight. A haunted house. Boys disappear around the corner into another room. I see them pulling thin legs out of saggy black underwear. I stand still. There's no way I'm getting naked in front of older boys. I was smaller than a gumdrop, I'm sure, and an uncircumcised Pointy, too. That's what Ethan used to call me, the only uncut boy among our friends. I'd call him Roundy, but it didn't have the same effect.

"Off your shirt, off your pants," Directreesa says, the landscape of clasps and wires of the large bra visible beneath her terrycloth short-sleeved shirt. "We must keep our hungry Amerikanyets clean."

I try to swallow, but my throat is dry as sandpaper, heart plummeting down an elevator shaft. I can't take off my clothes in front of some old lady and forty strange boys. A couple of them are staring over, waiting. Perhaps wondering whether the American might have a golden penis, bracing themselves to be blown away by my exotic genitalia. Directreesa isn't budging as if this is as routine as taking storeroom inventory. I slowly unzip my hoodie, hoping to stall for time, eyes searching for the exit. Directreesa puts down her clipboard and reaches for my shorts. I pull away. The only way out is forward. I unzip and let my frayed Levi's drop to the rough floor, my eyes following them. As bare as a little girl. I might as well have stepped out of flower print panties.

I walk, cold, to the rusty showerhead. A couple of older boys are lathered up, laughing, spitting water at each other. Feet stomping. I stare at the green tiles, the mildew in the cracks. I let my breath settle. The water taps at my chest. I look around. I feel shriveled and refuse to look down in case there's actually nothing there. Tears are fighting to well up. Until I notice the boy next to me. Uncircumcised, too. And another. And a third. Every-

where. Long, short, fat, teensy. I feel my legs loosen. I slow down
my scrubbing, heart settling. I take my time with the soap. I look
down. We are all Pointys, every last one of us. I breathe out.

At a kiosk outside the House of Clothes, a dusky-faced
man in a tracksuit rang me up a Disney DVD for Kosh's
daughter, a bouquet of I-Come-in-Peace flowers for Yulya, a
couple of bottles of Bavaria nonalcoholic beer, and two Cu-
ban cigars. I missed the shawl-clad babushkas with hennaed
hair and rubber thimbles on their thumbs. When I got to
Kosh's courtyard, I dialed him on his cell. He said he'd be
right down. I scanned the stairwell doors, guessing which one
he'd pop out of. I'd seen a few photos over the years and knew
Kosh shaved his head. He'd been going bald since his late
teens. For a while, he wore a woman's wig at his metal shows.
No mane, no cred. A few minutes later, a dark-clad figure
emerged. He wore black pants with a chain looping to his
back pocket, black T-shirt, black boots, black ball cap, black
jacket.

As Kosh bounced closer, I saw that he bore an uncanny re-
semblance to game show host Howie Mandel, the gleaming
egg-shaped head, the soul-patch strip down his chin. Well,
Howie Mandel after a rough night. Kosh's face was drained of
color and his forehead bumped with scar tissue. We hugged
and stared at one another for a few moments. It was good to
see him. "Hard to believe." He smiled, revealing that familiar
gap in his front teeth. "Been so long."

Kosh stepped back to take a look and was probably disappointed. No chains, no studs, no earrings or bracelets. I was
wearing jeans and a short-sleeved Western-style shirt from the
Gap. I'd traded my motorcycle boots for wide, round-toed
KEENs which I swore by for post-run recovery and comfort,
sweatbands instead of spiked bracelets. I wondered whether

he'd see my new healthy lifestyle as a betrayal. Not only had I left him and Kolya behind all those years ago, but I'd dropped the mantle, had gone straight.

"The Moscow Marathon?" Kosh asked. "My lord, sounds crazy. You run every day?"

"Four or five times a week," I said, then added, "Helps me see straight."

It didn't register.

"Tell me you have some cigarettes, Kalyeb," he said, eyes darting around the courtyard.

"I don't smoke anymore," I said. "But I brought a couple of cigars."

He looked quizzically at me and we sparked them up. "Do you inhale these things?" he asked, turning the stick over. "Tastes fierce. Oh, well."

Kosh reached into his jacket and handed me a CD. His first solo album, he said. On the cover, Kosh stood in white fog, edged with musical notes. Behind him was a bridge and cityscape, rendering him a giant rising from the mist. Same baggy pants, same wallet chain he was wearing today. A cross dangled around his neck, an open black hoodie. The album was called *Lyosha Korolyov: Signs.* I read the song list, which was in English: "Firestone," "Shadow," "Tears of the Rain," "Train." "It's like jazz metal," he said. "That's really what I want to play these days." I asked if metal was still popular in Moscow. "Of course," he said, incredulously. "We get tons of people at our shows. The whole city knows us."

I'd have liked to have given the disc a listen, but it seemed like he wasn't going to invite me up. I was secretly relieved. The courtyard felt more comfortable. Maybe this was the deal he'd made with himself, meeting me on the street. He wanted to maintain control. I respected that. In fact, it intensified the camaraderie I was feeling, both of us veterans of the bottle. I handed Kosh the DVD for his daughter and the flowers. We'd sit on the bench and puff fine cigars and sip our no-booze beer, two guys talking about the war and life away from the front. Sounded lovely, in fact.

"Foo, I can't smoke this," Kosh said, stubbing out the cigar and tossing it in the trash. "I think I have cigarettes in the apartment. Let's go. My wife's at work. Plus, my mom would love to lay eyes on you again."

I couldn't picture Kosh's mom at all. The last time I'd seen her, if you want to call it that, I'd been blitzed in a police station holding cell and had mistakenly called her a whore. I sure hoped she'd forgotten about that.

We walked into his warm, tenth-floor apartment. I untied my boots and slid into a pair of too-big cracked leather slippers from a pile kept by the door for guests. I stepped over a large German shepherd sleeping in the hall and into the living room, where a woman stood at the stove stirring a sizzling skillet. She turned around to say hello. Kosh's mother was short with dark hair. She still looked unfamiliar. We shook hands and she smiled. I was relieved. She glanced at the Bavaria beer I'd set down. "It's nonalcoholic, Ma." Kosh smiled.

"I don't know what to do with this one," she said to me. "Drinks too much. It's going to end badly."

Kosh was sweating. A floor fan swept the room like a searchlight. A small, adorable girl with a chestnut bob appeared in the doorway. Kosh's daughter, Katya. "Hi, Katya," I said. "I'm Kalyeb, a friend of your dad's." Kosh gave her the Disney disc and said, "Katya, go find a present for Kalyeb." She left and didn't come back.

Katya was a third-grader at School No. 80, the same school I'd gone to twenty-five years earlier. Where I'd learned to smoke and swear and trade in *Star Wars* figures and Levi's. Where, despite my best arguments, they wouldn't let me be a Pioneer ("It wouldn't be proper," the directress told me). I was the only student without the crimson neck scarf or Lenin pin, throat bare. I'd never received final grades, was always passed, wasn't learning much. My parents pulled me out midway through my third year after I'd started playing hooky. They stuck me in the American school, which insisted I drop back a year because my grasp of English grammar was so weak.

Kosh wanted to show me a video of one of his concerts on the computer in his bedroom. Katya's bunk bed stood in one corner with a desk and computer beneath it. The little girl had her hand on the mouse and was clicking away. Kosh leaned over her shoulder. "I'm doing homework," Katya whined. "It'll just take a couple of minutes, Katya," Kosh said, grabbing the mouse and clicking on a folder on the desktop. Katya left the room and two minutes later handed Kosh the phone. It was Yulya. "OK," Kosh said, hanging up. "I'll show you later, Kalyeb," he said, logging off. He headed to the bathroom, his third visit.

Kosh placed two plates of grilled chicken on the coffee table for lunch. He talked about his Land Rover, the Audi A6 that's next on his list, that he planned to celebrate his next birthday on Crete. After a while, Kosh's mom and Katya left for her French lessons. I waved. "Say goodbye to Kalyeb," Kosh instructed. She turned to her grandmother and the door shut behind them. Kosh put on one of his discs. "This one sold really well."

Kosh air-fingered an invisible fretboard along his forearm, a silent searing solo. I'd always loved that. As an American in Cold War Moscow, I could bring Nikes and Levi's and Black Sabbath tapes back from trips to England and Finland, but what I'd really wanted was to be Kosh. No matter how many spiked bracelets I wore or the number of Motley Crue and AC/DC patches my mom stitched on my jean jacket, his talent kept him above everyone, above me. Running the guitar pick up and down the strings, a blur of fingers dancing across the neck, he was a god creating lightning storms. He could produce that feeling I caught off a brain-scrambling Randy Rhoads solo or a crunching Jimmy Page riff, of making an insecure fourteen-year-old kid feel ten feet tall, capable of stepping over buildings. He taught me his chords and licks, but they never sounded quite the same under my clumsy fingers. I might as well have been wearing mittens. Two black Fenders rested in guitar stands near the armchair. I asked him to play me something. He shook his head. "The disc is better," he said. "I'm covering all the parts. Recorded it all right here." I was disappointed. It wasn't the same.

Kolya called from work, asking about dinner plans. Kosh begged us to pick a place in his neighborhood. "Kolya's is too far away," he said. "Getting home will be impossible."

"It's only five miles," I said. "What about the Land Rover?"

"It's in the garage. Pain to get to."

A cab?

"Both ways? They're so expensive."

The bus?

He looked at me as if I'd suggested he ride his balls home.

"Oh, lord, Kalyeb, I haven't been on one of those in twenty years."

I told him I had a few things to take care of back at Kolya's and we'd meet up later. Plus, I wanted to be alone for a couple of hours. Felt like I needed a break already. "I'll go with you," Kosh said, but then paused. "Is Olya there? She's kind of mean, don't you think?"

"She's home," I said immediately.

And that was the end of that. Kolya arranged to pick Kosh up on his way from work. As I headed out the door, Kosh said, "I'll have to have one *ryumka* tonight to celebrate your return."

"Be careful," I said. "You don't want to get in trouble at home."

"I told her that today was all about hanging with you, that I'd have to have a shot or two. It's been twenty-five fuckin' years after all."

"Fifteen actually, Kosh. I was here the summer of '95, remember?"

Riding the elevator, I wondered if I could have a single shot of vodka, maybe a beer to chase it. Would that be so bad? It would dispel the weirdness. Maybe I could connect better, maybe my Russian would come back. It hadn't been a proper reunion at all, certainly not like during my past visits when we started off with champagne and Stoli and stayed drunk for days, talking about all our adventures, blasting music, chasing girls. My sobriety was a wedge now. To turn down vodka was like saying you had two heads. To say you had a drinking problem, well, you might as well put on a dress. Beer wasn't even classified as taxable liquor here. So what would one drink, hell, even a good beer drunk, in another country, thousands of miles away from my life, really mean? What happened overseas stayed overseas, right? There'd been a lot worse drunks than me. I'd been sober almost as long as I'd been a drinker. That muscle memory was surely gone. I'd never promised anyone I'd never drink again. I didn't have an AA sponsor. I ran marathons now, for chrissakes. I'd paid my debt, right? My mind was getting tangled. I realized this was the first time I'd seriously considered a drink, where I was rationalizing. An important part of my sobriety had always been to return to Moscow and not drink. If I couldn't pass that test, I knew I wasn't in charge. It was probably why I hadn't been back in so long.

I thought about running the five miles back to Kolya's just to saw through this puzzle, to safely entertain a dark fantasy. Running was where I parsed anxieties and played out conflict scenarios—arms swinging like punches and blood pumping, ready to spill—to experience the emotions and sweat out any false desires. But I was in my marathon taper, taking a week off from the road before the start. I didn't dare deviate from the plan. I wanted my legs as fresh as possible. After Burlington, posting a sub-four felt well within reach. I crossed Leninsky Prospect near our old apartment building and waited for the number 4 trolley bus.

The Walters kids invite me over for a movie. They live in the American compound across the street and go to the American school. They have feathered hair and wear goose-down vests and their maid makes chocolate chip cookies from tubes of Nestlé dough sold at the embassy commissary. I know better than to show up in my Soviet school uniform and change into jeans and my Etonics sneakers. I race over. I love the sound of the metal canisters clashing together, the whir-click-click of the projector kicking on, the cone of light picking out the dust before the first images appear on the wall or screen. The promise of adventure, of escape. When I arrive, their projector is pointed out of their sixth-floor living room window and The Shining *is playing on the neighboring Soviet apartment building. The image is large and diffuse, but you can make out Jack Nicholson's face moving across windows and drainpipes. Everyone is laughing; Colonel Walters, drink in hand, looks pleased, and his kids at the open windows mesmerized. Within Nicholson's massive watery grin and crazy mountainous eyebrows, startled citizens appear on balconies and in windows like some live-action Advent calendar, trying to figure out what's going on. I know Russian kids who live in that building. Their mothers had served me tea and wafer biscuits. I slink back just in case they can see me.*

When Kosh and Kolya showed up a few hours later, Kosh was lit. That one-or-two-shots thing was obviously out the window. I remembered this all too well. Meaning to have only two or three drinks and waking up two days later surrounded by bottles and disarray, tongue pasted to the roof of my mouth and tiny carpenters, behind the fog, hammering two-by-fours to the inside of my skull. As Kosh unleashed a mighty stream in the toilet, Kolya said Yulya had found his hidden vodka stash. "He comes home drunk, and it's over," Kolya said. Kosh sauntered into the kitchen, an unlit cigarette between his lips. "Kolya, don't you have anything to drink?" Less than two feet away, behind a cupboard door, there were seven or eight bottles of duty-free Scotch and whiskey. I held my breath. I still wasn't sure how easily Kolya could be tipped. "We don't have anything, Kosh." Kolya smiled. "There's some nonalcoholic beer if you want."

"Foo," Kosh sneered.

I might as well not have been in the room.

We ended up at Sushi Planet, a Japanese franchise a few blocks up Leninsky Prospect. The second-floor dining room was all pleather booths, bright lights, and mirrored walls, creating a funhouse feel. The Slavic waitresses were dressed in kimonos, their dyed black hair tied back in buns and pierced with chopsticks. Olya lit a cigarette. Kosh scanned the menu and we ordered beers (nonalcoholic for me). "Guys, let's go somewhere else," he said. "There's no vodka. What kind of place is this? How can we celebrate properly?" Kosh called a server over, who confirmed his nightmare.

"Can I bring my own?"

"With a receipt," she answered. "And we have to serve it."

"Kolya, come with me," he said, grinding out his cigarette. Kolya winked at me and they left.

They were back within ten minutes. As Kosh talked with the waitress, Kolya said that as soon as they left the liquor store, Kosh chugged a third of the bottle. Olya shot Kolya a look. A few minutes later, a chilled glass bulb of Kosh's vodka appeared on the table, along with several beakers. I was heartened when Kosh didn't press any on me. He raised his glass and threw it back with such force I thought he'd swallowed it, too.

The drunker Kosh got, the less he spoke to me, the less I was there. He talked over Kolya. I wished I was anywhere but here. My return was just a reason to get wasted. This had nothing to do with me. I actually felt a flash of hatred for Kosh. But then again, how many times had I found the thinnest excuse to party? Some meager accomplishment, a visit from an old friend, even getting fired. It's really a selfish act, driven by the central desire to get fucked up, to escape who you are rather than to enjoy who you're with.

As a former drunk, wasn't I supposed to understand, to feel for Kosh? Just six months earlier I'd published an essay for *Runner's World* about running as a sobriety tool and how it dealt with the ghosts and demons, how it reminded me of important lessons like one foot in front of the other and accepting my limitations. Runners and former addicts from all over e-mailed me, some for advice or to echo my experience, others for help for a family member. As I had with them, I wanted to tell Kosh how once I'd gotten sober, running had opened up new doors in my mind and my heart, how I felt capable at long last, how it even got me high. That those first dry years, hiding away, scared, fat, depressed, my blood still stagnant in my tire-shaped body, the nightmares still streaming, the phantom hangovers, the panic and the anger, all that can be overcome. But writing a thoughtful response to a stranger was one thing, finding myself in the middle of the slop another. It's not as if I could tell Kosh to just strap on a pair of New Balances.

I was on edge, my confusion growing. It was as if I'd never lived here, never once got wasted with the guy. I felt a total stranger. I took a sip of my Bavaria Zero and glanced at Kolya, who was watching me as if he could see the gears turning. He lifted his glass in a silent toast. I felt better. Kosh whispered loudly to Kolya, "Kalyeb looks old, don't you think?"

The sushi arrived in a wooden Viking boat complete with mast. Little egg-studded rolls huddled in the center like mini slaves, a heap of ginger and mountain of wasabi on the bow. The cherry tomatoes mystified Kosh. During the meal, he had six or seven shots and unraveled at an alarming pace. He snapped for the "geishas" and then forgot what he wanted. He pressed Kolya to escort him to the bathroom. I wondered if part of me was jealous. Was he living the dream all of us once shared together? He didn't have to worry about work attire, putting in for vacation time, or even showing up. He got paid for doing what he was born good at. Strangers loved him. I thought about the fog he stood in on his album cover and wondered if it was more than symbolic. That it just wasn't clear where Kosh had arrived. I felt left behind and glad not to be him at the same time.

"I don't understand," Kosh said, picking up some unseen

thread of conversation. "I gave her a child. I bring in money. But that's not enough."

His cell rang and he stared at the number.

"God, who is this now?" He looked again at the screen. "Allo?"

He hung up. A few minutes later, another call.

"Kolya, who the hell keeps calling me? Allo, what do you want? Allo? Allo!"

He was getting angry, then scared. I wondered if his drunken fingers were missing the answer button. He had another shot. The phone rang again. This time, he didn't answer it. I wondered if it was Yulya.

"Hold on, who are you people?" Kosh blurted, as Olya guided a Philadelphia roll into Kolya's mouth.

Kosh wiped his face with both hands as if fingering away mud from his eyes. "What station is this? I don't know you people."

He was flat-out hallucinating. I had no idea what was going to happen next, but I was sure it wouldn't be good. Other patrons were turning in their booths.

"It's us," Kolya said. "Olya, Kolya, Kalyeb. Where are you?"

"I don't know you," Kosh answered.

"You don't know where you are?" Olya said, laughing. "My lord, Kolya."

"I know you," Kosh said to her. "But I don't know these other people. Tell them to stop looking at me. I have to get out of here."

Was it me? Was I the one freaking him out? Kalyeb with shorn hair, no cigarette burning between his fingers, a nonalcoholic beer nestled in his palm, being quiet, even judgmental. Was I as upsetting to him as he was to me, both of us reminders of who we were and who we'd become, the deep caverns that time had carved into us no longer echoing? Kosh slid from the booth. Kolya grabbed him by the arm. We asked for the bill. Outside of Sushi Planet, Kosh listed forward; the ground wanted him badly. Kolya propped him up. "This is how it goes with Kosh now," he said to me.

"They hardly see each other anymore," Olya said. "Can you blame him?"

Kolya called Yulya and begged her to let Kosh come home.

"He was drunk when he showed up," he explained. "Kalyeb doesn't drink at all. I drank beer. Olya's with us."

Long silence.

"But he's got nowhere to go. It's not a good idea for him to sleep on the streets."

More silence. Then Kolya hung up. He hailed a gypsy cab and put Kosh in the front seat. The driver buckled him in. Kolya handed him some bills and gave Kosh's address. I tried to grab Kosh's hand and say goodbye before the door closed, but his eyes were blank, head lolling. Who knew where he was? As I watched the car speed away, swallowed up by the dark boulevard, I was glad to see Kosh go. I felt bad but was relieved. I had reached some sort of mile marker I couldn't see. I watched Kosh become a pair of red car lights and let him slip away.

Boris and I head off to buy cigarettes. I'm dressed in my baby-shit-brown Soviet tracksuit pants and white T-shirt. I'd begged Mom to let me wear her black Casio digital watch today, to impress my friends. It has a smoked gray screen and buttons on every corner. As precious as jewelry. Boris and I argue about who looks older. I'm thirteen, he's fourteen. Legal age is sixteen, for cigarettes and booze. His voice is already deep, but he's shorter. "Ladno, I'll try," I say. We step around the corner from the ki-

osk to count our change, my heart pounding with nervousness. Two men approach. The taller one carries a briefcase, as if on his way from work. The shorter man is stocky with a pockmarked face. "Say, rebyata," *the tall man starts, "can you spare some change for ice cream?" I answer, "Sorry, we barely have enough for ourselves." Then I think maybe he could buy the cigarettes for us and am about to ask when the tall man looks around and says, "I don't think you understand; we need some money for ice cream." I'm startled at the brusque change in tone. I gulp hard. "Um, how much do you need? I can spare maybe thirty kopecks." He thinks for a moment, lets an old couple loaded down with net bags stuffed with vegetables waddle past, then says, pointing to my wrist, "I need as much as that watch costs." I look at my wrist for a second and am relieved. "Oh, well, I don't have that much, sorry," I say, thinking that's the end of it, and start to step away. "OK, we'll take the watch, then." Then it breaks over me, and I'm frozen, my legs rooted to the pavement. "Or we take you around the corner,* dadim tebye pizdi, *a beating you'll never forget, and take the watch anyway. Your choice." I look at Boris, who has a blank look on his face, waiting to see what I'll do. I feel like I have no way out. Terror has blurred all my senses. I can't even gulp. I slowly unbuckle the strap, hold out the watch, and look away. It's gone before I turn back, and so are they, melted into the crowd. "Why didn't you say you were an American, Kalyeb? Speak some English?" Boris snaps, as if I've shamed the both of us. "I'm going to round up the guys, we'll find those* pederazi." *I start across the avenue toward home but don't make it a block before bursting into tears. Mom is going to kill me.*

A noon start in mid-September felt late. It was 65 degrees and partly cloudy, pretty mild already. Most fall races back home got under way first thing in the morning, a cool upper 40s or low 50s. But no one seemed concerned. Giddy participants were streaming down Ulitsa Il'inka toward Red Square, numbers snapping against their chests, their images reflected in the glass storefronts of GUM—a onetime massive Soviet department store turned high-end mall, gilded with Louis Vuitton shops, Prada, Armani, and Apple. At the other end of Red Square, past Lenin's red granite tomb, tourists were getting their pictures taken with Stalin and Gorbachev impersonators.

Behind St. Basil's Cathedral, a cluster of white tents was set up near an enormous construction fence around where the Hotel Rossiya used to stand, once the world's largest hotel with 3,200 rooms, back when the Soviet Union was all about the biggest. It had been demolished to make way for an entertainment complex and underground parking garage. A lithe woman in bright fluorescent briefs and jog bra stretched out in front of a lanky teen in a shiny gold basketball uniform, much to his delight. A couple of thin bearded old men with white undershirts tucked into wrinkled cotton shorts kicked their feet up at their palms. They looked like they'd just run in from the collective farm. Listening to them talk about their paces and goal times in Russian, I felt trilingual.

An inflatable arch, marking the start, was set up on Vassilevski Spusk (St. Basil's Slope) behind St. Basil's Cathedral, perhaps Russia's most iconic image, with its swirling candy-striped onion domes. Legend has it that Ivan the Terrible (in Russian it's Ivan the Awe-Inspiring) blinded the architect so that he could never replicate his masterpiece. Some four hundred years later, in 1987, a West German teenager named Mathias Rust landed a twin-engine Cessna on Vassilevski Spusk, evading detection by the mighty Soviet air force. The humiliation allowed Soviet leader Mikhail Gorbachev to remove top military brass and expedite his political reforms, which unwittingly led to the collapse of the Soviet Union four years later.

The crowd of two thousand gathered near the starting arch. Half were running the 10K. Not a Kenyan in sight. No corrals or pace placards, either. A few grizzled men in cloth caps stretched out, touching their toes, twisting left and right, figures straight out of a Soviet calisthenics manual from the 1950s. Their calves looked like they were carved from wood, and they wore hard leather sneakers with thin black socks. You could be sure they'd never seen the inside of a modern gym with its trainers and gleaming equipment and weight contraptions strung together with cables and pulleys. They probably hoisted farm machinery and tree stumps and the weight of their own bodies to keep fit.

Nearby, a tanned blond in spandex leggings and a belly shirt swayed as if in a club, white ear buds wedged into multiple-pierced ears. A jewel glinted from her belly button. There were so many eyes on her, she might as well have been dressed in peacock feathers. Not one but two older runners wore shiny Lycra weightlifter's unitards. Three doughy middle-aged women in thick makeup, purple hoodies, and bright pink tights laughed with each other. Sculpted dudes decorated with tribal tattoos jumped up and down. One runner wore nothing but a grass skirt. The only costume I'd spotted. It made me smile to see a Russian drawing attention to himself like that.

In fact, they all looked so Western, as if they'd been stripped, sanded, and repainted. Cold War Moscow had been all grays and browns and reds, not a lot of blush and mascara and eyebrow waxing. America, too, had changed over the years, but the bends and twists were clearer in the rearview.

Here, it seemed all right turns. I saw flickers of the place I used to live—a cigarette with a cardboard filter, thick square eyeglasses, a Tolstoy-style beard, the warm rush of the air from the subway tunnel—but the sensations were like lightning bugs. Gone as soon as I caught sight of them. I squeezed in toward the middle but kept changing places, forward a few paces, back a few feet. Nothing felt right. Finally, I settled on the side.

We're crowded at an imaginary line between two orange cones, beneath a bridge on the embankment road. The gray Ukraina Hotel, with jagged spires, looms over the Moscow River. The embassy has started staging 5K and 10K races. There are about seven adults, including Pa. Mike D., the son of a consular officer at the embassy, and I are the only kids. I'm embarrassed by Pa's outfit, yellow shorts and a blue singlet with roses stamped on it, clouds of armpit hair puffing out. Mike and I are dressed in Adidas shorts and shirts. He let me borrow his red pair, even with the built-in underwear.

Mike would be the first person I ever get drunk with, the following New Year's. His will also be the first friendship altered because of booze, his parents forbidding us to hang out after we slaughtered his father's liquor cabinet and stumbled about on the frigid, snowy streets in shorts and T-shirts, singing Asia songs. A dirty correspondent's kid. We've signed up as a lark. We're both twelve. He's a swimmer, with a ripped six-pack that has me green with envy. I'm a six-workout-a-week gymnast and can whip off thirty pull-ups no problem. There's no way I won't cream Pa. A whistle blows and we're off. Mike and I sprint, chatting and laughing. It's quiet, a Sunday, light traffic on the leafy boulevard. At the 5K mark, past the government guesthouses behind a tall stone wall, Mike gets a stitch in his side and falls back. I stop with him, catching my breath before finishing the long climb

*to the Moscow University overlook and the ski jump that hurtles
athletes toward the city all year round and onto a stretch of hay.
Then Pa passes me. I can't believe it. He doesn't say anything
but looks like he wants to, his gaze straight ahead. I watch him
get smaller. A diplomat runner passes. I take a couple more deep
breaths and run full-tilt boogie, passing the middle-aged man,
with his sweatband and digital watch, then catch up to Pa and
haul ass right by him, not even looking at him.* Who, me, old
man? *I reach the top of the hill and the flat overlook, the construction cranes, Olympic stadium below. Then I feel a sharp
pain below my stomach and slow down, stopping in front of the
Metro station. I hide in the crowd as Pa passes. Again, I watch
him. The finish line is about two kilometers away. I figure I have
a minute or so. When my breath fully returns, I pour on a sprint.
Faster and faster, arms slicing. Pa becomes bigger and bigger, but
I still have a ways to catch up. Finally, I pull within twenty feet,
but then he's across the finish line, a small crowd of spouses and
kids clapping and handing out drinks, followed fifteen seconds
later by me. Out of breath and red-faced. Last time I ever run a
stupid race, I swear.*

Cheers went up and the crowd surged. A Revolutionary-
era military brass band played the runners forward. A stray
dog was curled up in the corner of a bus shelter, sleeping. We
poured across the six-lane bridge and over the river. A giant
neon Nikon sign crowned a large nineteenth-century apartment building across the way. The pack wheeled down onto
Bolshaya Ordynka, where we were met with pre-Revolutionary factories and apartment buildings straight out of *Doctor
Zhivago.* I half expected to see Bolsheviks hanging red banners from the ornate windows declaring the building communal property. But that bubble was pierced by a green Starbucks
sign. We passed a bed-and-breakfast with a brass nameplate

where a couple of tourists with suitcases at their feet watched the flood of flesh and running shoes. As far as I could tell, the race had not been advertised with banners or commercials. I didn't see any TV cameras. There were no barricades to hold back the crowds.

While the course has morphed and shifted over the decades, the Moscow Marathon has absorbed the sweat of some interesting times: Brezhnev's long era of stagnation, Gorbachev's glasnost and perestroika, Boris Yeltsin's populist fury and the collapse of the Soviet Union, the rise of the oligarchs, and the looming shadow of Vladimir Putin. Today, the route starts behind St. Basil's, crosses the Moscow River, and loops around Bolotnaya Square, then back across the bridge toward Vassilevski Spusk. Runners then make a ninety-degree turn and drop down to the embankment road. From there, it's four circuits along the river, punctuated by the long, massive walls of the Kremlin, a fortified citadel dating back to the 1300s, complete with cathedrals, palaces, and the presidential residence.

We turned right onto Bolotnaya Ulitsa and looped around the park. The leaders had already crossed back over to the Kremlin side and were trucking along the embankment road. The low stone wall across the water blocked out their legs, so from our side of the river, it looked like they were gliding on Segways. A couple of tourist boats plowed through the gray water. Within minutes, we wheeled back up onto the bridge where a few clusters of fans cheered—one girl wore full-size angel wings—and the rumpled band was still playing and the dog was still sleeping.

The sun had melted the clouds and warm rays bathed the course, boosting the mercury into the 70s. A runner passed me with an image of Jesus pinned to her chest, another with a

shirt touting membership in ultra-nationalist Vladimir Zhiri-
novsky's party. I saw no charity runners. My bib number flut-
tered in the breeze, printed on cheap paper hardly thicker
than newsprint. On one side of the road stood pudgy militiamen; on the other, bored-looking soldiers. I wondered if any
celebrities or political heavies were running. In any case, no
crazed fans were going to bum rush the course. In fact, within
a mile from the bridge, there were no spectators at all, as if
they'd been vacuumed up like crumbs.

The first drink station was barren. Boxes of bottles sat under the tables and the cups were stacked in plastic sleeves.
Runners were stopping but being turned away. I later learned
that only marathoners were to get water, so to avoid the 10K
runners drinking up the supply, nothing was put out until after their finish line. No one complained. Rationing was part
of the Russian DNA. I almost smiled. A tall mustachioed
man in wispy blue shorts and a straw hat dashed by me in
bare feet, nothing between him and Moscow. When I looked
at his pink heels and smooth arches, my Brooks Beasts felt
like Moon Boots.

A few miles later, a disheveled man staggered down from a
side street, wearing loose, wrinkled workman pants, his arms
hanging like a scarecrow's. He looked about sixty-five, could
have been forty for all I knew. These were the drunks I remember. It was 12:30 and the guy had no idea where he was.
The cops pushed him away, laughing, and he shuffled back
up the street, the walking dead. The sight of that trembling,
disoriented Soviet-style boozer actually brought a measure of
comfort, another lightning bug. But his bulb stayed lit just
long enough to absorb the light. As I ground across the warm
asphalt and through the cool shadows beneath the bridge

archways, I felt bits and pieces of old Moscow reconstituting, seeping into the rubber beneath each foot.

The trolley bus is packed and smells like sweat and onion and wool. I'm wearing my thin tracksuit, on my way to gymnastics. The only American ever accepted into a Russian program. My teammates are all younger, having started when they were four and five years old. At eleven, they are ripped and cut and whip off double back flips on the floor no problem. They train twice a day, six days a week. I train only in the afternoons. At the next few stops, more people get on; no one gets off. We're pressed into one another. At first, I think it's just hands and hips shifting, but after a few moments there's no mistaking it. Someone is fondling me, hands squeezing my crotch. My arms are pinned, my voice is pinned. I'm so shocked, I hardly believe it's happening. The hands keep kneading. I can't move. I keep saying nothing, terrified to meet the eyes of the offender. The groping persists. I look up. It's an old man in a brown fur hat, grizzled face, a pensioner. He doesn't look at me but it's clearly him. I can see his shoulder moving. He stares ahead as if nothing's happening. No one is aware. My voice box is suffocated, my heart thrashing. The doors open. It's my stop, thank God, and I push and claw my way out. I don't even look at the bus as it pulls away. I just keep my gaze down and head straight for the gym, past the kiosks and through the apartment courtyard. I've never felt so helpless. And alone. I hate myself.

At the next station, tables were spread with cups, but each filled to only a couple of fingers of water. Not an electrolyte in sight. I grabbed two, spilling some down my face and chest, splashing my bib. Within minutes the corners of my thin paper number began flapping, the edges disintegrating, until I had to remove it. I folded it like some ancient document and slipped it into my pocket. The four safety pins left behind gave me a pierced punk look, reminding me of the cross earrings and spiked belts I later wore on the streets here, a metalhead in a land where Beatles albums weren't even sold, Michael J. Fox in *Back to the Future.*

At the turnaround point for the first circuit, which was shorter than the remaining three, I had to shout my number, *pyatdisyats devats,* "fifty-nine," to the race officials who were recording passing runners by hand. Confused foreigners were flashing fingers to the doughy women in official fluorescent vests sitting in folding chairs. As we headed into the second pass, numbers were littering the ground. This didn't bode well, but dealing with the unexpected—bad weather, cramps, blisters, dead batteries—was part of this game.

I strode past the Cathedral of Christ the Savior, which had been rebuilt in the 1990s after Stalin demolished the original in the thirties to make way for the world's largest outdoor swimming pool. The church's golden domes gleamed, warming my eyes. In the distance, a bouquet of glass skyscrapers twinkled and shimmered, balancing out Moscow's collection of toothy ministerial buildings. Below me, the Moscow River was like a medieval moat, heavy water held in check by a thick stone embankment. Not like the gentle Charles that separates Cambridge and Boston, hemmed by leafy dirt paths along its low banks, where geese graze and dogs frolic. Back home, I sometimes kept pace with the eight-oared racing shells, the coxswains urging me on, too. A river was my favorite running backdrop, the mysterious sliding current reinforcing my sense of momentum, of the journey, that desire for the sea, for the wide open. I had no idea where the Moscow River lets out.

Over the years, whenever people learned that I'd lived in Cold War Moscow, they said, "The formative years, eh. What was *that* like?" I always struggled with the answer, because it wasn't like anything. There was nothing I could compare it to, especially after the collapse of the Soviet Union in 1991. I might as well have lived on the moon. "It was interesting," I'd say with an arch of my eyebrows. But how could I sum it up? Moscow had been a life of extremes. Among my Soviet classmates, I'd been a star from the start. What had been run-of-the-mill back in DC was suddenly exotic on Leninsky Prospect. My Izod tennis shirts and Velcro sneakers were covered in stares. A *Star Wars* action figure was gold bullion. My Walkman flat-out mind-blowing. I was invited to all the birthday parties. For the first time in my life, I had swagger—no longer the smallest kid in class with big ears who smelled vaguely of last night's urine. Suddenly, I didn't have to try, a child's dream.

That is, until I was among the American diplomat kids, who cycled into Moscow on two-year stints, fresh from Virginia or Frankfurt or Bahrain, with all the latest clothes and gadgets and catch phrases from Hollywood blockbusters. Hamburgers, pepperoni pizza, Doritos, Snickers were all at their fingertips, things I had access to just a few years earlier. Correspondents' families were allowed to shop only the dairy section of the embassy commissary (Soviet milk soured quickly, often on store shelves) and to eat at the American-grill-style snack bar only during certain hours. Among the dip kids, I spoke more softly and tried twice as hard. For a while, I told my Soviet friends I had a twin brother who went to the American school and didn't speak Russian, that my parents were trying an experiment. That way I could ignore their shouts of *"Privet, Kalyeb!"* from across the street when I was with my American friends. The last thing I wanted was those two worlds colliding. No one bought it.

My parents embodied this feeling of shuttling between

two poles. Pa was highly educated, from an aristocratic pre-Revolutionary family that lived all over the world. Mom was a country girl from northern England, raised on Ouija boards and ghosts by an eccentric, tough-talking, Oxford-dropout mother. There was no money for college. Pa had covered the U.S. State Department and the White House, flew on *Air Force One* with Nixon on his famous 1974 China trip. Mom freelanced stories about divining rods and haunted houses. She battled her weight and was always dieting, while my dad was skinny and ate ravenously. Mom was gregarious, outspoken, confrontational. Pa was quiet, serious, and almost humorless, at least around me. An editor of his once commented: "Part of his structure is having a straight backbone. He lives by his own code. . . . Face tightly drawn against the world." Both their childhoods were marked by the disruptions of World War II, and they bonded over the early loss of a parent. When Mom was twelve, her father died of lung cancer. Pa's mom was taken by an aneurysm when he was fourteen. Mom humanized him; Pa legitimized her.

I sometimes wondered about the path my life would have taken if we'd stayed in DC. The clock would have kept ticking. Would I have kept wetting the bed? Would that have gotten my ass kicked in the public junior high school? Would I have been studious and straight-laced like my older sister? Might my love of comics have led me toward graphic design or my growing collection of rare Daredevil issues have implanted in me a sense for investing and business or antiques? Would my obsession with D&D have turned me into a video game developer or fantasy bookstore manager?

After our return from Moscow, whenever I beat myself up for not understanding something ("Who's Eddie Murphy?") or some horrible social faux pas (blowing pot smoke in someone's dorm room, not knowing to use the hit towel) or just had a general feeling of not wanting to be with myself, I found comfort by subtracting five years from my age or joking, "Yeah, there's that black hole again." But just as the clock on my development was set to resume, the bottle drowned it out again, the ticking always far offstage. By my thirtieth birthday, I had grown a hell of a lot younger. Maybe running was my way of speeding things back up, living as much of my life as possible at six and seven miles an hour. A multiplier, finally.

By the time I'd completed my second pass, my mind was starting to numb. The long red Kremlin walls with their sniper slits, the river, the embankment parks, the ornate bridge underpasses were all pleasing to the eye on the way out. But then the scene, to be absorbed a total of five more times, was like being waterboarded by architecture. And the two lanes of runners—outgoing and incoming—were a constant, will-sapping reminder of how many people were ahead of me. Only in Moscow would a course be designed like that.

A few miles later, the road became eerie quiet. Nothing around me but swinging arms and pumping legs. The collection of shoes pounding the asphalt sounded like golf applause. I felt marooned in motion. My sense of the course was confused, distance and time leached of meaning. Since runners covered the same route multiple times, a variety of mile-markers dotted the road. You'd have a 5 followed by a 35 followed by a 20 followed by a 10. All in kilometers. There were no course clocks. I gave myself a headache trying to figure out the kilometer-mile conversion, then dividing by my time. I didn't even think all those skipped math classes would have helped. I guessed an hour and fifty-nine minutes at the halfway mark.

This mental exercise reminded me of the Terrible Math. After a long night or a bender, I tried to total the ounces consumed before I'd passed out. Budweiser came in all sizes, from 8-ounce pony bottles to 16-ounce pounders to the forty; shots had to be translated into 12-ounce increments and wine always threw me off. Anything under 96 ounces and I felt safe, none of that existential paralysis and self-loathing. Anything over 120 and I'd probably been an asshole. For years, I relied on this calculation for a sense of how I should feel about myself. Though I didn't realize it until later, the sub-four, the calories burned, my pace-per-mile, the readout on the bathroom scale, had assumed a similar role. I looked for my reflection in those digits. But instead, the numbers only teased an addiction-prone personality to the surface.

At the next fluid station, I noticed a tray of black bread and a mess of salt. The Russian version of energy goo? What the hell, I grabbed a chunk, dipped it in the salt pile, and popped it in my mouth. It was, well, salty. Then it was really salty. My mouth started swelling. I felt I had sausages for lips, a New York socialite jogging Central Park. Jesus Christ. I ripped a bottle of Powerade from my fuel belt and drained the whole thing. Where the hell am I? I slid the container back in its holster and wiped my mouth.

I'm back home from Northfield Mount Hermon. It's the first winter break and I landed at Sheremyetevo Airport a few hours earlier. In the tongue of my Nike hightop is an eighth of pot, still wrapped in coffee grounds and sprayed with Polo cologne. Kosh, Kolya, and I are in a stairwell in a far corner of my apartment block. It's great to be back. We're drinking wine and vodka and smoking cigarettes and listening to Led Zeppelin on my ghetto blaster. Kolya takes a swig from the bottle and throws his arm around me. "When the Levee Breaks" bounces off the tiled floors and painted stucco walls, Robert Plant's train engine harmonica and high, wailing voice. We're singing along, the booze draining through us. Outside, snow is coming down under the anemic lights in the dark courtyard. I'm about to surprise them with the weed when the elevator clanks to a stop on the floor above. Two militiamen step out and come down the stairs. From below, another pair comes up the steps, snow dripping off their thick woolen boots and long grey coats. "Nu ka, rebyata, shto eto takoye? What's going on here, boys?" We're stunned, surrounded by police, who look all business. One steps over and shuts off the boom box. We hop off the window sill and put out our cigarettes beneath our shoes. "Just trying to stay warm," Kosh says. "We'll be on our way."

The superior militiaman glares at Kosh. "Don't you have any respect for the good citizens living in this building, playing this filth and filling the floors with smoke and spit? Get your things, let's go. Now."

My heart begins pounding through the booze and I'm scared to speak. I glance at Kolya, who has a flat look on his face. As they herd us out of the stairwell toward their police jeeps, Kolya whispers to me, "After that car, run. Get the hell out of here, Kalyeb." I think about it for a second, touched that Kolya wants to protect me, but I can't make a move. I'd like to think it was because I don't want to leave Kosh and Kolya to their fates, that we were in this together, but I'm scared the militiamen will give chase, catch me, squirming in their thick hands, and then it'd be worse. Even drunk, I'm afraid to act. They pile us into the back seat of a small jeep, with me forced onto Kolya's lap. Sitting taller, I feel a boozy courage well up and start yelling that I'm an American and that if they don't let us go, they'll cause an international incident.

*But my passport is at home. "And where did you learn such good
Russian,* Amerikanyets?*" one says, dubiously. I switch to English,
and demand to know where we were going, but the words feel
self-conscious in my mouth. I jump back to Russian. No one looks
at me. "Keep your mouth shut back there,* maleesh. *We'll sort everything out at the station."*

*At the precinct, they sit us on a long bench near the hold-
ing cell, where a few grizzled drunks slouch against the wall, one
laid out on the floor. The bars are made of rebar and fuzz before
my eyes. The superior officer takes down our names, our parents
names, and phone numbers. During the ten-minute ride, all the
vodka catches up and the room is blurring at the edges. I stand,
throat already open. "Robots, you're all robots," I yell. I stagger
up to some kind of frame on the wall and spit. The place goes silent. Next thing I know my arm is twisted behind my back and
my face pushed to wall. "Nu ka, veterai!" The beefy militiaman
orders me to clean it. Behind the glass is a portrait of Vladimir
Illich Lenin, with his goatee and piercing gaze, the father of the
Soviet Union. I wipe the glass with my fingers, turn back to the
guard, and put them dramatically in my mouth. He shakes his
head and glares. I wonder if he's going to hit me. Kosh muffles a
nervous laugh.*

*Even drunk, a kernel of reality cuts through the haze. I've de-
luded myself into thinking I'm being brave for my friends, showing them that the system they lived in could be confronted. It's easy
to be a rebel when there are no consequences. Life could become
unpleasant for them and their families in ways I knew nothing
about. It didn't occur to me until later to wonder if Kolya had
wanted me to run so they wouldn't have been arrested with an
American. We were strictly KGB territory. Maybe he was just being practical, surviving.*

I don't know how much time passes. A woman shows up with a militiaman, hurries to the window and begins speaking. She has long brown hair, an overcoat and leather boots. Her shoulders are wet with snow. I can't hear what she's saying, but she isn't told to take a seat. She looks back at us, but her face is beyond my comprehension, a pan of water beneath a cloud of bangs. I'm annoyed that she seems to be cutting in line and yell, "Why do the whores get taken care of before us? This is an outrage." It was Kosh's mom.

It was clear that night that my future was already forking away from the dreary and dangerously static one that lay ahead for Kosh and Kolya. Of course, I always knew there were differences between us, and obviously, they knew, too. But we managed to transcend them. But not this. They felt the same way about the system, felt it more deeply, but never in a million years would they have acted the way I had. We were pried apart that night. That's when it really began. And it felt like my fault, my arrogance, my belief that I was someone I wasn't.

Under the next bridge, I saw a giant Gatorade container at the water station. My heart glowed. I'd looked all over Moscow for Gatorade. But nothing. The city boasted everything from Maseratis to IKEA to lab-grade cocaine, but no Thirst Quencher. As I drew closer, my eyes filled with refreshing visions: an icy orange waterfall dumped over a victorious coach; Michael Jordan switching the ball in his hands beneath the basket; close-ups of a football player's face covered in antifreeze-colored sweat. I noticed the cups were filled with a brown liquid. Maybe Moscow was a test market for a new flavor. Root beer? Caramel? I grabbed one off the table and gulped at it like a shipwrecked sailor. It was hot tea. I nearly gagged at first, but then it almost felt comforting. It tasted like the jostle of a train car, a warm kitchen, a joke that unfolds like a story. And I downed the rest.

Two and a half hours in and I felt like I was dragging a station wagon. The dreaded thought of walking seeped into my head. *Just slow down for a few minutes. People die every year running these things, you know.* I tried to snuff out the notion, but the image of myself walking persisted, the relief it would bring. *But if you walk even a single step, you can't claim you "ran" this marathon.* This was followed by questioning my very identity as a runner, my commitment. *It's OK, you don't have the will for marathons. Stick to your five or six miles. Or maybe find a different passion.* Devil voices. They knew the raw nerves. My years in Moscow had embedded in me a feeling of incompletion, not seeing things through to the end. Interrupting myself became a theme: failing classes, randomly breaking up with girlfriends, getting fired, getting expelled; everything half-assed. A state of suspension that had come to feel normal.

A few years after I got sober, I'd begun railing against interruptions. I never missed a story deadline. Being late killed me, even to a movie. The smallest break in routine, even the whiff of a hint that I couldn't deliver on a promise reinforced a sinking feeling of blackness, and destiny. I drove through blizzards to get to work, the only guy in the office, pools of melted snow at my feet. Soon, this extended to the run. I'd rather to piss myself than veer off course to look for a bathroom. Unless my tongue was bloated with thirst, I'd pass every water fountain. The worst was when I stopped my watch at a red light and forgot to start it again. My entire run was thrown off, minutes, pace seconds stolen. A paused run somehow counted less. I counted less.

Continuous flow was the reminder that I was no longer that fragmented, damaged person who blew off work, family, and friends. It was probably what drew me to marathons. A chance for an emphatic completion. I didn't want to be that helpless, aimless kid ever again. I still resented him. I pressed on: through his second birthplace, through his first graveyard, legs as heavy as anvils.

I glanced across the river at the Lenin Hills, since renamed the Sparrow Hills, the highest point in the city. Spilling from just below Moscow State University, the hills comprised a forested slope, cut with winding walking paths, a popular place for outdoor enthusiasts. Pa often met his sources there, away from listening devices and prying eyes. That's where he was arrested, where our time in Moscow came to an abrupt halt, all of us thrust into the middle of an international Cold War scandal. Accused of espionage, Pa was thrown in the notorious Lefortova Prison, held as a political pawn after a Soviet spy was nabbed in New York.

Mom whips open the curtains in the waiting room and knocks on the pane. Suddenly, I see cameras and lights and a crowd of faces in the window, aimed at us, at me. A guard rushes over and tugs the curtains back. Nelyzya! *he says, with a stern look. Another guard steps out from the table. "The press must report the truth," Mom says. "Not you, not even Gorbachev is going to stop them." My face flushes red. I'm embarrassed by how loud she is, by her thick accent, her determined fleshy face. The guard is unmoved and returns to his station. Mom starts to make her way back to the curtains when we are called up.*

When I walk into the visiting room, Pa stands from the couch. He's unshaven, wearing a grey cardigan, and his pants sag. No belt or shoelaces. He looks small, like a plucked bird. We hug.

I can't remember the last time we'd touched. The silver-toothed interrogator looks at my long hair and grins at Pa, "I thought you said you had a son, not a daughter. You have twenty minutes." A translator sits next to my dad with a notebook, monitoring the conversation. I ask about his cell, his roommate, the food, whether he can exercise. He's an interview subject, a stubborn mathematical theorem I've been assigned to figure out.

I think about Pa as an anxious bookworm of a kid, bouncing from Michigan to Venezuela to Paris to New Hampshire, under the disappointed glare of an aristocratic Russian exile father whose thick accent kept him out of America's elite country clubs. Even Pa's prestigious school degrees didn't win much affection from Serge. After finding out about Pa's first tour in Moscow for UPI in the 1960s, Serge warned, "You'll be arrested, tossed in the salt mines, and your passport won't help you." I wonder if Pa is glad Serge isn't alive right now. Or if he misses him.

When it's time to go, the interrogator turns to me and says, "Tell your mother to leave the press at home next time. They're not helping your father's situation." Something rises in me. Maybe it's the residue of the "daughter" comment, but I turn and glare, "You've arrested an innocent person. This is her husband. These are his colleagues. What did you expect—silence? This is their job." We stare at each other a moment. I feel weird, focused. Then the interrogator turns and smiles at Pa. "Well, what do you know, Nikolai Sergeyvich," he says, addressing Pa in the traditional Russian patronymic, "your son, he has a real Moscow accent."

My pace had slowed considerably. I was at three hours and twenty-five minutes, with six miles to go. I'd be lucky to break four and a half hours. I told myself it was the jet lag, the oily Soviet-style pasta I had the night before, the heavy drag of nostalgia. But it wasn't just me. By the last pass, the course was littered with the walking wounded. It was a flat-out death march.

On the opposite side of the road a young guy, with thick arms, was muttering like a wounded gorilla, still on the third circuit. At least I wasn't him. As he got closer, I could hear *"Nyet bolya, nyet bolya."* No pain, no pain. Yep, it was a full-out mind game now. I picked up his mantra and leaned forward. *"Nyet bolya,"* I said out loud. *"Nyet bolya."* My voice in Russian was deeper, vibrating off my tonsils; the words didn't dance at the front of my mouth like English.

Nyet bolya. I thought about Shea and Chris. I would run the rest of the race for them; that would keep me going. *Nyet bolya.* I pretended tragedy had struck and I had to run for help four towns away. *Nyet bolya.* Kosh's face stepped forward in my mind, his gap-toothed grin. I wondered if his marriage would survive, how he would parent his daughter, how his grasp on his circumstances seemed so tenuous yet he kept landing on his feet. *Nyet bolya.* My heavy footfalls were sounding off like sledgehammers dropped against the pavement. Each step became a *nyet,* a "no" to the monotony, a "no" to the pain, the "no" I should have said to every drink I shouldn't have had. *"Nyet, nyet, nyet."*

At the final fluid station, there was no water. "It's coming," the volunteer told me. "Run to the turnaround and it'll be set up by the time you get back." Deflated, I could barely gulp, spit thick as frosting, my throat coated in sand. In the final, carcass-dragging miles of a marathon, I found slaking my thirst almost impossible. As soon as I tossed a crumpled cup to the road, I was parched, scanning the horizon for the next drink table.

As the pedestrians stopped to watch the runners pass by, to stare at the spectacle and comment to one another, I realized I'd been chasing another feeling that Moscow had instilled in me—the sense of being special, the way I'd been gawked at, the ease with which I'd attracted Russian friends and girl-friends. That our apartment and phones were bugged only re-inforced the notion that we were worth paying attention to. Of course, Pa's arrest and the international media attention, the TV appearances and press conferences, launched me again into a special tier when we came back home, this time among my American classmates and teachers at boarding school. I got second, third, and fourth chances. I tried to play the at-tention down but made sure the Reagan-visit picture was al-ways displayed somewhere in my room. And as the story and attention faded, the booze, the drugs, helped me ease the pain of coming down off the high of unearned celebration.

In Moscow, you could get everything now; all the doors were open, and everything was on display. It was clear as ever that it had been the Soviet context, that oppressive, bloody framework, that had given me status. Perhaps that's why I'd been afraid of coming back sober all these years, stripped of my drunken artifice, afraid of being exposed as average in the New Russia, in front of Kosh and Kolya, in my own eyes. But a marathon regulated things, taught me humility. For these four-some hours, I was confined, by my body, to who I was, not who I wanted to be, or pretended to be. There was no bullshitting 26.2 miles. Running had taught me to exist in the middle. I could feel the pain of this effort, this earning of place. My mind perked up and my shoes felt lighter.

As I strained to determine the distance to that last turn-around, I knew that returning to Russia was more reconciliation than sobriety test. Even though the Moscow that had very much shaped me was very much gone, my DNA still swirled in this air like summer *pukh*. Through the sweat and labored breathing and screaming muscles the strand twisted clearer than ever before: the thirteen-year-old mugged for his Deep Purple and Def Leppard pins in broad daylight by older teens, the sneaky fourteen-year-old slipping 100-ruble notes from his mom's purse, the insensitive fifteen-year-old celebrating the death of another Soviet leader in front of the embarrassed family maid. My limbs loosened. Inside, I withdrew my fist, unclenched my fingers. I'd just been a kid, a stupid, sensitive kid going through changes in a foreign country. The kicks and punches I'd been throwing all these years had been traveling through nothing but empty space, never connecting. Dispensing punishment and actually inflicting it are two different things, and suddenly neither held much appeal. I only had the strength and energy to navigate the present moment. That was all that mattered.

When I got back to the drinks table, there were three plastic cups left. Two runners were in front of me. I snatched the very last one. I drank and tossed the cup. Then I stopped and turned around. And walked back to the table, dodging oncoming runners. I picked up the cup and wiped the rim on the driest part of my shirt and put it back for the next dried-up pair of lips. Then turned around. I forced one leg forward, then the next, willing my arms to give me momentum.

At last, I saw St. Basil's and its sweet, almost edible-looking bulbs. Delirious, I could have dived headfirst into the asphalt. I was actually going to get this thing done. I turned up alongside the wide bridge, the final uphill stretch onto Vassilevski Spusk and the finish. Up ahead, the course's only clock read 4:28:07. I recalibrated one last time and aimed to finish under 4:30. I crossed through the inflatable Asics arch with seconds to spare.

Still grimacing, pain stabbing my lower back, my quads filled with rocks, I shuffled through the roped-off chute. A moon-faced girl with thick pigtails handed me a bottle of water and a bag of mini muffins. I uncapped the bottle. *"Mozhno vtoroyu?"* I asked. "Could I get a second bottle?" Before she could answer, a gruff race official barked: *"Nyet, ne mozhno!* We barely have enough for everyone here and you're asking for a second. Keep moving." I was too exhausted to say anything. I wondered if I'd have gotten a different result if I'd asked in English. Russians weren't about elevating each other. "See a piece of grass growing taller than the rest, cut it down," was how the saying went. I almost smiled. I'd caught up to Moscow at last, chased down that part of me. This place would always live inside me; it always had. I shuffled off to a cluster of kiosks behind St. Basil's for water and got in line behind all the other thirsty Russian runners, the sun drying our glowing, salty, sweat-covered faces.

Kosh hammers the nail into the barrel, his long dark hair swinging into the weak beam of streetlight. He wedges the nail out, releasing the liquid. He captures the silver stream in an empty Pepsi bottle. I'm at the ready with another. After filling three containers, Kosh, Kolya, and I climb back over the dry cleaner's wall and head for the nearest stairwell. We settle on a wide windowsill between floors. Kosh leans his guitar in the corner and Kolya passes around the vodka bottle. I'm wearing Kosh's brown pleather vest that smells like his BO. I'm leaving for boarding school the day after tomorrow. I don't know when I'll see them again, whether we'll be allowed to come back.

Mom and I got in a fight a few hours earlier. She didn't want me on the streets, afraid I might be taken next. But there was no way I wasn't going. "Tvoi otyets matros," Kolya says, pulling a rag from his pocket, his thick legs encased in acid-washed jeans. "Your dad's a sailor, a tough SOB." I dampen my strip of cloth with the cleaning fluid and put it to my face, inhaling the sweet chemical fumes deep into my lungs. My eyeballs soften, my brain dancing with pricks of light, edges darkening. I've never thought of Pa this way, a tough son of a bitch. He doesn't like violent movies, won't even step on an ant. He wears highwaters, sports thick Woody Allen–style glasses, hardly ever raises his voice. How does that make him a matros? I don't understand. Does Kolya see something I don't? Does he have the answer? I look at him, his shoulder-length brown hair the same color as mine, but his eyes are rolled back, mouth agape, rag hanging between his fingers. Then I'm gone, too.

The next day, my legs were tight, knees aching like someone had used them for batting practice. I was thankful for the long Metro escalator. I was still disappointed in my time. I thought for sure Moscow was where I'd break four hours. A flat course and all that East Bloc nostalgia.

I pushed through the exit doors, into the fading sunlight. Ksenya said she'd be wearing a pink sweater. Kiosks and moneychangers surrounded the Chistye Prudy station. A plump, long-haired woman in ripped jeans swayed and twisted to music blasting from unseen speakers. Several other women walked back and forth with sandwich boards reading SPEAK ANY LANGUAGE. Car horns filled the warm evening air. Chistye Prudy was a popular spot for demonstrations, and metal detectors and barricades had been erected on the dirt lane that divided the boulevard. I hadn't seen Ksenya in at least twenty years but I recognized her immediately. Short, dirty blond, impish face. *"Zdorova, Kalyeb! Tak preyatno tebye vidyats,"* she shouted, as we hugged. "Great to see you, too," I said.

I'd been good friends with Ksenya's younger brother Kiril, who was in my class at School No. 80. I hadn't kept in touch with either of them after leaving Moscow. Last year, Ksenya e-mailed me out of the blue when she was visiting the States for work. She wrote that she was eight years sober, but Kiril was in bad shape. That he hadn't worked in more than ten years and had burned himself out of two apartments, passing out with a cigarette burning between his fingers. He was either drunk, in the hospital, or too sick to go anywhere. "I called him and he said he may try and meet us, he'd like to see you," she said, stepping back.

As school chums, Kiril and I had gotten drunk—and sick—a few times on my parents' Grand Marnier. He was one of the few kids I ever had a fistfight with; well, it was more like a wrestling match down a flight of stairs. We must have been thirteen or fourteen. I had no clue what started it. I think I may have, feeling the need to prove I could do more than flash my passport. I didn't expect him to fight back so hard. He slammed me against the sill, punched me in the chin. He was shorter than me but practiced Sambo, a Soviet martial art, and quickly had my feet out from under me. I kneed him in the stomach and grabbed at his hair. Our friend Boris just watched, then finally broke it up. I was breathing hard, jaw throbbing, trying like hell not to cry. Kiril spat on the floor, called me a bastard, and split. He didn't give a shit that I was an American. In fact, that he didn't hold any punches back told me he'd always seen me as an equal, a friend.

Ksenya said she didn't visit her brother anymore because

he was too aggressive and unpredictable. They only talked on the phone or spoke face-to-face when she saw him at the hospital, where he periodically landed. Ksenya said she once tricked Kiril into an AA meeting, but he left after fifteen minutes. "He doesn't like talking about himself and doesn't like hearing other people talk about themselves." I was nervous to see him. Part of me hoped he wouldn't show up.

Ksenya's own road had been rough. She became a drunk, too. She told me she lost two husbands, one to a drug overdose and the other to suicide. Her stepdaughter was killed in a car accident. It was just her and her twenty-four-year-old daughter now. Amazingly, she still had that same sunny disposition I remembered, an easy gap-toothed smile. She was a firm believer in AA and looked, for the most part, healthy. Her yellow, slightly rotten teeth were the only visible traces from her previous life.

"There's Kiril," Ksenya said.

A group of guys was walking toward us. I couldn't pick him out. Then he appeared, a half-smile on his face. He still looked short, but there was less of him. He wore a faded Gap ball cap, a white tennis shirt, loose jeans, and a brown leather jacket. *"Privet, Kiril,"* I said, as we drew into a hug.

"Vot eto da," he said. "Well, this is something."

His voice, which was always a little raspy, had dropped a couple of octaves. It sounded like his throat was coated in bark. I looked at his cheeks, drained of color. His nose was smudged as if singed and the corner of his left eye had a pink growth in it. Small scars ran above his eyebrows. I couldn't even see the old face in there. But his eyes were still the pale blue I remembered. We stepped back to take each other in.

"You look well," he said.

I wanted to say the same, but he'd know I was lying.

We grabbed McDonald's and brought it back to a bench along the lane. Youths were gathered on other benches, bottles of beer and wine at their feet. Kiril pulled apart a newspaper and laid down sections for us to sit on. Then he reached into his pocket and gave me a 10-ruble jubilee coin and a Soviet-era emblem from a military hat. He pulled out a chocolate bar for Ksenya. I was moved. Ksenya gave me a worn rubber key chain. The objects felt priceless in my hands. I felt awful I'd showed up empty-handed and didn't know what to say. I'd lost that instinct. Kiril lit a cigarette.

"Your dad still alive?"

"Yes."

"Thank God," he said, wincing on the filter, his hands etched with scrapes and healed cuts.

Kiril slurred and wobbled a little, as if he had been drinking, but I didn't smell anything. Ksenya said he'd been sober a few weeks but was still sick and planning to go into the hospital. He pulled his hat off, revealing rows of thin hair, as if to say, *This is who I am now.* His teeth were worse than Ksenya's. Several were missing; others looked black. He told me his apartment had no light, no TV. "I don't even have toilet paper to write on. It's like a barn." He said he had an eighteen-year-old daughter he hadn't seen since she was two.

"You don't shave," he said, apropos of nothing.

I hadn't in a couple of days, but it wasn't by any stretch a beard. His cheeks were smooth and I wondered if he'd cleaned up for the occasion. I felt bad for not recognizing the significance of this meeting; guilty for my health, the privileges of my passport, for being able to pop across the ocean to do something as frivolous as run a marathon, when Kiril could hardly crawl out of his dank apartment.

Ksenya invited us to her AA meeting down the road and we made our way. The meeting was being held on Sretenka Street in a conference room belonging to the Moscow Helsinki Group, a human rights monitoring organization. Ksenya introduced me to the group leader, Katya, a reporter at *Novaya Gazeta*. "One of ours who survived," Ksenya told me.

Ksenya ran to get me a twenty-four-hour chip as a souvenir. It didn't occur to me until that moment that my own anniversary had once again passed without notice: eleven years. It bothered me when this happened. In fact, September 9 was the day I'd boarded the plane for Moscow. I reminded myself not to read any significance into the timing or the geography. Symbolism is OK for literary critics but dangerous for drunkards.

The room was humid, packed mostly with men, mostly middle-aged. Kiril and I sat next to each other in a back corner. He took his hat off and rested it on his knee. The meeting was themed around the first and ninth steps: that we are powerless over alcohol and our lives have become unmanageable, and making direct amends. The session opened with a communal reading of all twelve steps and the twelve traditions. A sheet was passed around the table and members took turns reading. Introductions were made. I was surprised when Kiril said, "Kiril, *alkogolik*." I followed, "Kalyeb, *alkogolik*." It was the first time I'd ever uttered that statement in a recovery setting, to a group of strangers. It didn't affect me at all; I might as well be ordering soup. I never liked the clinical nature of the word *alcoholic*—seemed too clean and clipped for such a messy, painful, and sprawling state of being. And I was never comfortable with the *ing* in *recovering*. But at the end of the day, what's in a name? We were related, all of us.

Between readings, people told stories about their final

bender, how the program had saved them, why they kept coming back. Some of it I didn't understand. It was strange to hear the language of recovery and shame expressed in Russian. It seemed jarring, like the radio stations in Olya's car that played the exact same music as in the States, with Russian DJs making similar banter and jokes, a blueprint laid over another culture.

Kiril glanced at me a few times, pointing to the sobriety chips on the table. "Do I get a prize for going into the hospital?" he said, with a grin. He nudged me when a thick woman squeezed her way into the room, a Soviet-era throwback in a split skirt, mannish hands, and hennaed hair. It was as if we were back in grade school. He was clearly not into the scene. When talk turned to alcoholism as a psychological disease, he whispered to me, "It's a problem of the soul, which means only God can solve it." Then he patted me on the leg, stood, and left the room. That was it; he was gone, the last time I'd see him, I figured. Well, at least he'd lasted this long. I understood the discomfort, the running away, the not wanting to belong.

I respected people who attended AA—anyone, for that matter, who had chosen to swear off the bottle. No doubt, the twelve-step model has given comfort and hope to millions of addicts navigating the dry, arduous landscape of sobriety. There is hard-earned wisdom in AA meeting rooms. Along the way, I have culled bits and pieces, embracing one of its mantras: "Take what you need and leave the rest." So why didn't I join their ranks?

It wasn't a deliberate move, more like naive inaction. I felt so stunted, as small and shriveled as a tequila worm, that it would have been impossible to open up to a room full of

strangers. I couldn't define myself as a recovering alcoholic before I could determine who I was, or at least get a few good
glimpses. I knew striking out on my own might slow down
recovery and there would be mistakes along the way, but I
hadn't known my true self since I was a teenager, not since we
moved here. I needed to give myself a chance.

I was also wary of groups and well-worn paths and people who claimed to have the answer. Psychiatrists, psychologists, chemical dependency counselors, and people in recovery programs the world over are constantly debating the best
approach. With talk therapy? Spirituality? Pills? A combination? Neuroscientists have entered the fray, searching for the
chemical cause and solution. The questions seem to multiply.
Is there an addiction gene that gets passed down like diabetes?
Is it a disease that happens to us like cancer? Is it an acquired
physical dependency that spreads to the mind? A manifestation of underlying organic issues—depression, panic, anxiety?
There was an array of ideas and views swirling around a simple activity, or nonactivity—not lifting the glass, not uncap-
ping the beer, moving through the next hour, the rest of the
day.

While some folks are uncomfortable with the religious
tones in AA rooms, and separation-of-church-and-state law-
suits have been brought against court-ordered AA sentences,
accessing a spiritual side of yourself, in my view, is an essential component for successful recovery — in other words, tap-
ping into something inside, something bigger than yourself,
slightly mysterious, that is beyond judgment and isn't fully
knowable but can be felt. Achieving a mindfulness without
thinking. Faith, humility, submission, amends-making, moral
inventories are indeed necessary ingredients in this effort. AA

can be a good place to start, but the recipe isn't exclusive to it. Spiritual conversion is too personal, and it had been happening for eons before Bill W. was born.

In a way, the multitude of recovery philosophies reminded me of the myriad thoughts on running, another simple endeavor. One foot forward, then the next, swing the arms, try to keep the head up, the back straight. Repeat, and repeat again. That's about it. Go to the end of the block, down to the river, to the next town—you decide. But for such a basic set of mechanics, the activity also generates a lot of discussion and opinions in the running and medical community: whether dirt really is easier on the joints than pavement, whether marathoning will ruin your knees, the benefits of barefoot running, the best way to swing your arms. *Runner's World* even devoted ink to the best way to tie your shoelaces. One of my favorite contributions to the conversation was a study out of Europe showing that nonalcoholic beer aids in recovery from marathon training.

Every time I see a new study or theory or prerace power food, I know I should want to figure out how to run better, to fuel better, but at the end of the day, I'd rather just run. I need simplicity. That's running's fundamental appeal. I'd lived in confusion for too long. Every time I run, I'm having a conversation with my purest self, my moral inventory on full display. Running is a state of being more than a sport, a way of life. Sure, people have entered recovery programs or AA rooms years, even decades, after their last drink. Time isn't always an absolute healer. As for me, I'm not ruling anything out in the future, but for now my path of recovery feels good and solid.

At the end of the meeting, a sack went around for money. I dropped in 20 rubles. "Kiril's gone," Ksenya said. "He does this every time." We gathered our things and headed outside, back to the Metro. In the corner of the courtyard, a cigarette glowed in the dark. It was Kiril, squatting against the wall. I was stunned, uplifted. "I needed some air," he said.

Walking back to the Metro, Ksenya tried to get Kiril to understand the concept of a higher power. "You can be your own higher power," I said. Ksenya was silent and Kiril looked at me incredulously. "What are you saying, that I'm higher than God?" I didn't have an answer. Sometimes it's best to keep your coping philosophies private, I thought. They're so individual as to be intimate. But as we talked, I could feel my Russian coming back, more sentences springing to my tongue, abstract concepts popping from my lips. It never took long for the words to wake, the more I hung with native speakers, with my people. But I couldn't find the words for: "Maybe God expects us all to simply act on his behalf, that's what I think."

In the subway station, Kiril and I exchanged addresses and I promised to send him my dad's book about his arrest, which had been published in Russian. I watched him shuffle off to his connecting train, ghostlike, his jacket hanging off him as if from a clothes rack. Sober steps all of them. Ksenya seemed hopeful. "Maybe the clinic will work this time," she said. "Maybe there will be a miracle."

4 hours, 29 minutes, 44 seconds
Average pace: 10:19
578th place out of 940 finishers

Gill, Massachusetts

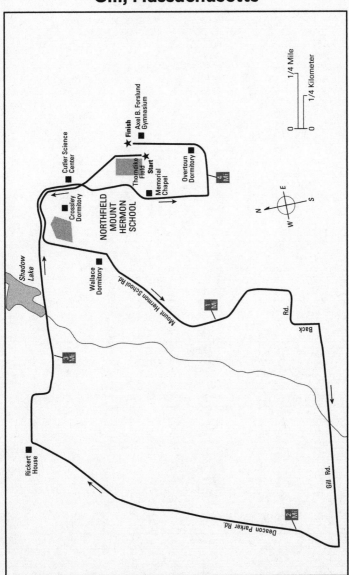

4

119th Bemis-Forslund Pie Race
(4.3 miles)

Sunday, October 18, 2009

"Your work to date in Religion has not been very inspiring or impressive. I sense some skepticism on your part about the course and wish you would express those feelings more openly to me, if this is the case. You also need to improve your record on getting to class on time." — Northfield Mount Hermon school chaplain Ginny Brooks, Religious Studies, October 22, 1985

"Caleb, I am sorry your rooming arrangement hasn't panned out better, since it was something that you wanted badly at the time. We need to work together to get you and Mike to be more civil. Mike has been more positive lately so there is considerable reason to hope for improvement between you two, if you try too." — Vaughn Ausman, floor parent, third floor, Overtoun dormitory, October 25, 1985

The rain had turned to snow as Route 2 lifted us into the higher elevations of northwestern Massachusetts, our first

taste of winter. It was mid-October. The dashboard temperature gauge dropped further as we got closer to campus, and
clouds of static drifted across the radio frequencies. I hoped
the weather would keep the turnout low. I wanted to get
in and get out. Northfield Mount Hermon, a two-campus
boarding school in rural Massachusetts, was my first American home after Moscow, and I'd long ago written it off, along
with the confused teenager that I'd been, trying to reclaim his
birthright.

The giant gold, blue, and maroon sign with the lamp-and-
flame logo that long dominated the brick-columned entrance
was gone. The new signage was more modest and subtle,
butter-colored with blue lettering. The long access road unrolled through tall woods, past a few faculty houses, and past
Shadow Lake, where the seniors battled the juniors in a massive tug-of-war over the water every year since 1926, one of the
school's many grand traditions. Somewhere in these woods on
graduation day twenty-one years ago, Pa walked our Jack Rus-
sell terrier, trying to compose himself, tears coating his cheeks,
knowing he still had to give the keynote address while I sat in
the infirmary choking down my final major school rule violation, my expulsion papers being drawn up by the dean. I'd
driven through campus a couple of times over the years when
we'd visit Chris's parents in nearby Northfield but never got
out of the car. Never went to reunions. Never saw my name
in the alumni notes section of the school magazine. But now
I was back to take part in one of NMH's most venerable and
celebrated annual events: the 119th Bemis-Forslund Pie Race,
a 4.3-mile cross-country scramble, by some accounts the oldest continuously held footrace in the world. At stake, a ten-
inch apple pie.

The biggest change at NMH was that the school had

shrunk in half, one part of its identity cleaved away. With the school facing financial woes, the Northfield campus, two miles to the north, had been sold to the C. S. Lewis Foundation, which planned to establish a religious college there in 2012. New administration, arts, and admissions buildings had risen on the remaining Mount Hermon campus, where I'd laid my head for three years, but the place looked pretty much the same, only tighter.

Chris sat in the passenger seat, staring out the window, running shoes on her feet. She'd been a day student here six years before I arrived. She never went to reunions either. I steered the car past Wallace dormitory and the tennis courts. There was West Hall, with its massive columns and hair-netted cooks; and the Cottages, cozy hilltop dormitories that overlook the football field. There was the post office and student center. And Beveridge Hall and Holbrook Hall, where I met every week with Associate Dean David Schockett, a requirement of my near-constant disciplinary probation status, his soul-burning eyes studying me from above a bushy beard, long fingernails on one guitar-playing hand—in my mind, the devil with elbow patches, De Niro peeling an egg in *Angel Heart*. My surrogate father.

I pulled into the parking lot in front of the Axel Forslund Gymnasium, pointing the car at the football field, the stone chapel looming on the hill, where the entire school gathered for announcements and community and song. I stilled the wipers and stepped out. There was a slight drizzle on the wind. Maintenance men were setting up the finish area, roping off the finishers' chute and putting up a tent. I didn't see any other runners. I was glad. Last thing I wanted was to bump into anyone who might recognize me. This context in which I'd lived still harbored some of the same emotional cues like how you revert to the dumb, dependent, sniveling kid around your parents. Without a mirror, I could believe I even looked the same—mop of brown hair, bangs covering shifty hazel eyes, tight, uneasy smile, skinny. I blinked hard a couple of times. A thin yellow stripe marked the finish line.

The Pie Race was preceded by the Mooney Tart Race, a short dash for children and fac brats, named after Pat and Ginny Mooney, the longtime track and cross-country coaches. The bushy-haired Pat Mooney coached the NMH runners when I was a student. In fact, I ended up taking cross-country running as a PE with him because nothing else fit my schedule. I usually cut from the trail after a quarter-mile and made my way to Blake Student Center for an egg-bacon-cheese muffin. I'm sure Mooney could tell from my dry forehead that I wasn't running the full course. Even though he never caught me, he kept me on his radar.

The next semester I made the varsity Ultimate Frisbee team. Mooney spied me a few times at the outdoor smoker behind Beveridge Hall. Cigarettes were forbidden if you played on a team and he chastised me, ordering me to 'fess up to my coach. Sheepishly, I said I would, then didn't. A week later, Mooney cut behind me in the cafeteria line at West Hall, decked out in blue windpants and a gray x-country T-shirt, and tapped me on the shoulder. "Did you tell your coach?" I said, "No." "Then I'm going to." I turned to him and, because I was with some guys from the dorm, said, "Go ahead, then. What's stopping you?" Then I turned to the dessert rack and grabbed two butterscotch puddings. My friends "oohed" and "aahed" me, but I felt shitty showing my back like that. I failed varsity athletics that semester. Even though I didn't get one of those carpeted NMH letter patches like the rest of my teammates, I ordered a school jacket anyway. I had my name stitched on the chest and "Varsity Ultimate Frisbee" on the back and took my seat back at the smoker behind Beveridge Hall. I puffed countless cigarettes in that maroon, leather-sleeved jacket, a disdain for runners flickering like a pilot light in some tiny corner of my heart.

"On Sunday, Nov. 10, dorm faculty discovered a liquor bottle in your room. At the time, the bottle contained hair conditioner, but because possession of such alcohol paraphernalia is a violation of a major school rule, you were referred to me . . . you are being placed on Disciplinary Probation for two terms . . .

"In our initial discussions about this incident and about other school rule violations you may have engaged in, you were initially less than forthcoming about your behavior. When you decided that you could be more honest, you spoke about a series of behaviors that are both clearly violations of school rules and actions that are simply not in your own best interest. You described several incidents both of drinking and of smoking marijuana on at least six different occasions between the middle of September and the beginning of November . . .

"I would ask you to think seriously about the choices you made; a clearer understanding of yourself and your motivations will be essential if you are to successfully stop this and other behavior that is potentially damaging to you." — David Schockett, associate dean, Mount Hermon campus, November 25, 1985

I grabbed my bag from the car and headed for the locker rooms. I pushed through the heavy swinging door and into a yellow-tiled cavern. The rows of lockers and yawning emptiness conjured a sense of nakedness and menace. I changed next to an older man, who introduced himself as class of '75. "I'm '88," I replied, feeling a bit like a liar since technically I hadn't graduated with my class. But how do you articulate an asterisk? I stared at him, fascinated that after all these years, someone would still feel such connection to his high school. I didn't get it. When I was done with something, I was done. The experience was severed and shelved away. A square peg in a square hole. Any revisiting would alter the past and affect the present, upset the order. Even in Middlebury, I'd lived less than a quarter-mile from the paper where I used to work, hung-over much of the time, and where Chris still laid out pages. I never visited. It would have been too surreal. Down the bench an even older runner with a scrub-brush beard tugged on running tights like wrinkled black skin. He looked like he'd been carved from driftwood and reminded me of the images of bushy-bearded NMH founder D. L. Moody himself.

Old Moody started the annual race in the late 1800s, shortly after founding the Mount Hermon School for Boys, "an earnestly Christian" institution. The Northfield campus was then called the Northfield Seminary for Young Ladies (the school would merge and go coed on both campuses in 1971). The evangelist set the race course for six miles and made participation mandatory, convinced that physical activity, the brain, and the soul were intertwined. "The hand, the head, and the heart" was Moody's motto for the school he founded, drummed into us every Wednesday at Chapel. Every student, regardless of economic background, had a work job—dreaded cafeteria duty at West Hall, the farm, the bakery, the laundry. At the time, they were meaningless words to me, an institutional slogan, school propaganda to be flushed down the toilet. But sitting on that bench, I recognized the connection between Moody's three elements. It was a continuous thread—not separate entities orbiting each other—that could be woven into wisdom. I was here after all, running to understand something in some earlier version of my heart, doing exactly what Moody had prescribed 130 years earlier. Had NMH planted this seed? In such barren soil? Was that possible?

NMH dates the Pie Race to 1891, which would make it six years older than the venerable Boston Marathon. The winner was given a medal. Otherwise, you ran for the glory of your dorm. In the early 1930s, the school's athletic director, Axel Forslund, shrank the course from 6 miles to 4.3, made participation optional, and convinced an alum named Henry Bemis to fund the medals. And he added another incentive: apple pies. "It was something different, something to eat and something to share with roommates," Forslund told *Sports Illustrated* in 1979. "Boys were always hungry."

The first three male and female students, and the first alumnus and alumna, received hardware. Everyone else was shooting for pie time—thirty-three minutes for males, forty minutes for females—and a ten-inch, two-crust, all-American apple treat made that morning in the school bakery. My problem was all my training had been for endurance, not speed. I was a solid nine-minute miler. I could probably bring it down to eight minutes, but I wasn't expecting to be wiping sugary glaze from my lips. I figured just showing up was worth something. I zipped my bag, wished the class of '75 guy luck, and shuffled back to the car to wait with Chris for the start. I watched the mist coat the windshield until everything outside was a blur.

"On the night of February 28, you spent time in a dorm room where alcohol was being consumed. This action constituted a violation of your Disciplinary Probation and, thus, rendered you liable for dismissal from NMH . . . the disciplinary consequences of your action were a four-day suspension and the extension of your Disciplinary Probation through to the interim of Fall 1986.

"Caleb, I am very concerned about you; you do not seem to understand what it is that is being asked of you. I do not hear you asking, 'What is it that I must do to show that I'm serious about being here?' Rather, you seem to be asking, 'What can I still get away with and not get kicked out of here?' You simply aren't taking responsibility for yourself. Do you need help doing so?

"Are you willing to put the effort into your academics—effort that seems to be lacking, according to most of your teachers. Are you willing to address honestly your fascination with the drug and alcohol culture here at NMH?

"It's time to start growing up, time to start taking some positive control over your life. Anything less, I'm afraid, will either result in a waste of time for you here or it will create conditions that will not allow you to remain." — David Schockett, associate dean, Mount Hermon campus, March 10, 1986

Some running genealogists trace America's first running boom of the 1970s back to NMH's Pie Race, specifically the 1963 edition. That's when a sixteen-year-old sophomore named Frank Shorter toed the line on a lark; he was a skier, a football and baseball player. He placed seventh, behind the five-man cross-country team and one cross-country skier. "That's what got me interested in running," Shorter later told a reporter. "That and the fact I was getting my rear end kicked all over the place playing football." He tried out for cross-country the next fall and won the Pie Race in 1965, establishing a course record that would hold for ten years. Seven years after that victory, he won marathon gold at the 1972 Munich Olympics, raising the profile of the sport to new heights. In his wake, hundreds of thousands of Americans laced up running shoes and took to the streets, including Pa. Thinking about that timeline that now ended with me here in my running shoes, I couldn't help but think about the concept of destiny, something I'd never put much stock in. I liked to think we control our own fates.

In 1979, Shorter returned to NMH, along with other alums, faculty, staff, and students, to run the Pie Race and regain the course record that had been broken four years earlier by a junior. Chris was a freshman at the time and remembers being called to the window at Cutler Science Center by her chemistry teacher to see Shorter streak by on a practice run. "It was a huge deal. All the TV cameras and *Sports Illustrated* were there. Everyone was buzzing about it. I didn't really know who he was, but he looked fast."

Sarah Pileggi covered the event for *Sports Illustrated*:

With Axel Forslund's gun they were off, a mass of color streaming past a background of bare sugar maples and a gray stone chapel on a hill; past Crossley Hall, where Shorter recalled having warmed his pies on a radiator when he was a student; past the tennis courts and Shadow Lake, where the hockey team used to play before it got an indoor rink and its own Zamboni. On a winding drive leading downhill toward the school's main gate, about three-quarters of a mile into the race, Shorter took the lead. Out the gate they went, onto a two-lane highway for a few hundred yards, then off again onto Turners Falls Road, a dirt lane through school property, with barren winter woods on one side and fields of corn stubble on the other.

At the halfway point, Shorter was flying and the pack was far behind. The runners made a left turn at Day's Corner, 2.7 miles out, around a pile of rusty farm machinery and onto North Cross Road, past a goat tethered in a side yard, past a colorless, weatherbeaten scarecrow and some beehives.

At three miles, the course turned left again and started up Overtoun Hill, a steep, quarter-mile pull, dotted with white frame faculty houses. From the top of the hill, with its view of the river and rolling wooded hills to the horizon, it was down to the playing field, around its south end and through a corridor of cheering onlookers to the finish line at the flagpole.

Shorter won, clocking a 20:54 and shaving a minute off the course record. "I really ran hard for the first time in a long time, and it felt good the whole way," he said, then posed for pictures with his pie and his fans.

"Caleb appears to have survived his brush with History with aplomb and good humor. Because of the flood of attention he was getting before his father's happy release, I intentionally held off, on the theory that he'd appreciate the psychological 'elbow room' and because, from what I could see, he was weathering it all quite well. Now that Caleb's life is returning to 'normal,' I hope to find occasion to know him better. Already I know how bright he is. I've encountered his droll wit. And to my immense satisfaction (and relief), I've not yet had to encounter any unacceptable decibel levels from his electronic musical gear."—Jerry Reneau, floor parent, fourth floor, Overtoun dormitory, October 12, 1986

"Are you still in this course?!"—Felicity Pool, Health teacher, Peer Education, October 15, 1986

As a student, I'd never heard of Shorter and thought the Pie Race was stupid, wouldn't have thought of entering, not for a moment. I was too busy learning how to fashion a pipe from a Skoal tin and figuring out whether I should like the Grateful Dead (assuming, at first, they were a metal band). At NMH, I was surrounded by real Americans. Not the dip kids of Moscow's foreign compounds, but bona fide genuine articles, having lived their whole lives in places like Austin, Texas; Columbia, South Carolina; New York City; Boston; and La Jolla, California. I was among rock stars. They were tribal Americans with their ripped and patched jeans, tie-dyed T-shirts, flak jackets, and Doc Martens. Their blood ran with stripes and stars and bits of cool. Mine was just red. After a few months, I hid my spiked leather belt and Mötley Crüe T-shirts at the back of my bottom drawer.

Before my dad's arrest, I was an obscure, skinny kid in the furthest dorm on campus, a perceived Soviet among Americans—solidified by my last name and the school, inexplicably, hanging the Hammer and Sickle among all the other international flags at Convocation Hall. I wanted people to like me, but even more than that, I didn't want people to dislike me. And if they didn't get to know me, they'd never have a chance. I both longed, and feared, to be different.

After Pa's arrest, the summer before my junior year, I was catapulted to celebrity, suddenly the best-known kid among the 1,200 students, the subject of crushes and dedications on the school radio. Reporters camped out at my dorm. I gave press conferences and fielded requests from Letterman and Oprah. It made certain things easier for sure—people wanted to get high with me, for starters—but it kept me in a different kind of outside. The American in Soviet school again.

"An event came to my attention which shed serious doubt on the appropriateness and maturity of both your judgment and behavior: I have spoken at length with the Dean of Residential Life and the Headmaster concerning your involvement with psychedelic mushrooms during interim break.

"I need to once again underscore the seriousness of the situation you have put yourself in: any subsequent violation of school rules or of your Disciplinary Probation will cause you to be raised for dismissal . . .

"I am requiring that you participate in a drug evaluation with one of the school's consulting psychologists. We will carefully

review together the report that is forthcoming and act on what-
ever recommendations may be offered as a way of helping you
deal with your relationship with drugs." — David Schockett, associate dean, Mount Hermon campus, November 11, 1986

"Clinical Impressions: Despite Caleb's own perception that he
does not have a substance abuse problem, the evaluation results
suggest his past usage of both marijuana and alcohol reflect a pat-
tern of abuse. It seems quite likely that if Caleb had not been dis-
covered in the mushroom incident, he would have continued reg-
ular use of alcohol and other chemicals . . .

"It's also possible that Caleb's unique lifestyle (moving alter-
nately from the U.S. to Moscow and back) has exaggerated his
interest in trying substances by placing him in one culture where
drugs are unavailable and mysterious and then transferring him
to another culture where the same drugs are available . . .

"Diagnosis: Alcohol Abuse, Episodic." — Donna K. Nagata,
PhD, licensed clinical psychologist, Greenfield, Massachusetts, November 23, 1986

As three o'clock drew closer, the crowd swelled to more
than a hundred, mostly students. The air was raw and the sky
knitted with gray clouds. Chris and I took our places at the
line chalked on the field, the same playing field where a news
helicopter landed to take me to the Hartford airport in exchange for an exclusive interview about my father's release.
Chris disliked races but agreed to run this one. She didn't like
timers and competitors. Chris ran without a watch and would
stop to help a turtle cross the road, watch a deer grazing at the
edge of the woods, or slow down through a lilac patch. After,
she'd ask me if I'd seen these things and I'd answer no. I never
stopped. "It was like running through grape soda," she'd say.
Even though I was faster and could run farther, it would take
me a long time to catch up to her as a runner.

I hadn't recognized any of the alumni names from the 1980s sign-in sheet and the older faces all looked like strangers. One student was dressed head to toe in a green body suit with goggles, an alien Gumby. Girls wore wispy shorts and form-fitting tops and tightened their ponytails with gloved fingers. Boys stretched and blew into their hands, jumping up and down in tight leggings. I was wearing baggy windpants, long-sleeved Under Armour, and a thick T-shirt, with a skullcap under a brimmed running hat. Compared to the students, I might as well have had a shawl across my lap to shut out the chill and an ear trumpet to hear the starter's pistol. These kids seemed so easy, all with the same relaxed bearing. If they were rule breakers or wracked with insecurity and anxieties, it didn't show.

An NMH communications staffer with a video camera was interviewing older runners, asking them about their fondest school memories. My heart started thumping and a twitch entered the back of my neck. I thought of my first panic attack when I was called to recite a poem in Mr. Hamilton's English class, a feeling of sheer terror that came out of nowhere as I left my seat. My fingers shook and my heart felt like it was going to hurtle out of my chest, voice warbling, everything a blur as I felt all my classmates' eyes boring holes through me. I thought I was losing my mind. Afterward at the smoker, I tried to play it cool. "Yeah, I think I had an acid flashback up there." But these attacks would follow me into college, grad school, editorial meetings, any gathering of more than two people, and would inform my hangovers, amplify every feeling of fear and cowardice. I was eventually prescribed pills to reduce the symptoms, but over the years, I learned to run before any presentation or meeting. An hour of sweaty roadwork felt just as effective at equalizing the pressure. I once ran in jeans, leather gloves, and wool scarf before a meeting out of town because I didn't have running clothes. Tearing up block after crowded block, I felt like a kid chasing after a school bus. As the woman and her camera got closer, Chris and I slipped to the back of the pack.

Then the gun went off and a mass of legs and arms and backs spread forward as if spilled from a bottle. The field was squishy, sucking at my running shoes. The bare calves in front of me were already speckled with dirt and bits of leaves. The herd stampeded past the library where I stole Nabokov books for my shelves — mostly to impress girls — and the Cutler Science Center, where, well, not much happened. We left the grass and poured onto the asphalt access road that ribboned around Crossley, the largest dorm on campus. I started to loosen as my body filled with movement, my mind infused with motion. I settled into the run, feeling the contours of the road, the wind stroking my cheeks, clearing myself for takeoff.

I kept a brisk pace, my shirt wrinkling against my chest, as I wheeled past Wallace dormitory. Gaggles of girls stood in the smoked windows, cheering and screaming, some holding signs I couldn't make out. I blew up a kiss. Swaddled in my layers, I probably looked like somebody's creepy uncle. Boys in shorts and muscle Ts streaked by me, all cuts and divots and bouncing hair. Gliding as if they were running on one of those moving walkways at airports. Surprisingly, their speed didn't bother me.

The road unrolled down a grassy hill and I let it carry me, catching my breath as my legs wheeled beneath me, teetering on the edge of losing control, feet almost outpacing my mind's ability to command them. I still liked that feeling, the deep end. I streaked through a swath of woods and past a white clapboard house. I wondered who lived there and realized how little of the campus I'd explored as a student. A few girls in NMH singlets whom I'd passed earlier regained their leads. But when I hit the first mile marker I was shocked. 7:11. I'd never run this fast. I didn't recognize myself.

"Your Disciplinary Probation Contract is a good one: I suggest you pin it up near your door so that you can review it every day as you leave your room."—David Schockett, associate dean, Mount Hermon campus, December 22, 1986

"An understanding of Caleb continues to elude me. Since I feel that Caleb prefers to elude understanding, I will not here engage in a desperate improvisation."—Jerry Reneau, floor parent, fourth floor, Overtoun dormitory, March 14, 1987

"Caleb seems not willing to use his mind, but rather in a passive attitude reaches some conclusions about what he is studying—at least so in English. He did show moments when he was perceptive in his comments; yet if anyone—including a classmate—asked him about a point he was making he quickly withdrew into silence . . . I suspect he is intellectually capable of much more; may he start using his mind."—English teacher Richard Fleck, Literature in Translation, March 21, 1988

I held steady and at mile 2, my watch read 7:09. Unbelievable. I might actually have a shot at this thing. At first, this was an uncomfortable feeling, the fact that I had an opportunity to capitalize. I was already testing my limits. To keep this pace for two more miles was daunting. It didn't help that I started cramping beneath my ribs. I lifted my arms overhead, trying to relieve the pinch, running as if I were looking for people to high-five.

The course pitched sharply, climbing, and leveled out at a farm field. I passed a few older runners. Were they faculty members? Had I sat in their back rows? I thought I recognized an old math teacher, a subject that had repeatedly twisted my arm up behind my back. Though the route had been modified over the years, I wondered if these were the same corn rows Frank Shorter burned past thirty years earlier. I followed a narrow muddy track, the potholes and divots filled with mulch. The path squiggled uphill and morphed into a rutted gravel trail. The terrain had changed. I was now greedily gulping the air with the urgency and crudeness of a burglar clearing a bureau top into a garbage bag.

I felt the speed drain out of me, pooling in my damp socks, as my feet picked their way around the divots and rocks, thighs thickening with effort. I was still plagued by Hard Starter's Syndrome. I always burst out strong, consuming everything in my path—whether it was steak sandwiches, ecstasy pills, or other runners—and gradually faded. I didn't have anything left at the end, no kick. Only the price tag. I'd miss the mark again at NMH. It would be appropriate, symbolic even. I was looking for a way out. Who cares about a stupid pie? Maybe it was naive to think I was a pie-caliber runner. I hadn't even calculated a pie pace beforehand, was just winging it. Exactly as I had when I'd lived here. I thought about that sixteen-year-old kid fumbling in the dark parking lot trying to let the air out of Dean Schockett's tires while he was in his office one evening. After the deed, I dashed back to the smoker and proudly showed my pal Jack the air caps I'd unscrewed. He burst out laughing. I hadn't realized you needed to push the needle with a nail or key to let the air out. I wasn't even good at being bad. I felt my pace dip, my arms grow heavier.

At mile marker 3, my watch read 8:30. I told myself to focus on the run, relax, one foot, then the other. The finish line will come, this will pass. I'd be back on the road to Beantown within thirty minutes, the fuck outta here. Maybe I just couldn't stomach conclusions. When I couldn't control outcomes, I resigned myself to their inevitability, good old-fashioned Russian fatalism embedded in my genes. Maybe we didn't control our fates after all.

But as soon as I recognized that old voice, the part of me that perhaps was only that seed twenty years ago cleared its throat. OK, now go to work, keep your eyes on the prize. Don't slack. I wasn't just an older version of that passive, alienated kid from Moscow fretting about his cool factor. I'd now run three marathons in six months. I could do this. All the training and hundreds of miles and blisters and chafing, the commitment, the routine of it all, somehow kept me pinned properly in place. I'd long ago traded vernaculars. Ounces referred to Gatorade, grams to saturated fats. Endurance no longer meant how long I could go without puking. Speed meant speed.

The last marker sign came into view, one mile to go. When I looked at my watch, I was surprised to see I had 7:34 left to make the thirty-three-minute cutoff. Holy shit. I'd already posted two sub-7:10 miles. I mustered what I had left and turned the key, forcing my legs into overdrive, and began running balls-out fugitive. If something was within reach, reach for it. Or else, what was the point? That was who I was now. Soon, I was passing dudes; they could see the focus in my gait, on my face. A different kind of pie-eyed. The thought made me smile. In front of the headmaster's house, where I'd gone after I was expelled to plead for an academic certificate since I'd taken my final exams, I saw that I had three minutes to get down the hill, around my old dorm, Overtoun, and to the finish. I blew a snot rocket and took the long hill like an escaped maniac, flying, above myself, a burning tire hurtling down a hill. Every step deliberate, only toes touching, the cold wind putting tears in my eyes. I kept them straight ahead, resisting the urge to look at my old room window. I needed every molecule of energy clocked in for the present.

"On June 15th, you left your dormitory after curfew without permission, a major school rule violation, your fourth. A campus-wide search was undertaken, much to the disruption and worry of fellow students and faculty. You left school grounds to spend the night with several other students and alumni at a motel in Northampton, where drinking and drug use took place. Given your history here at NMH, we cannot in good conscience find any alternative but to recommend to the headmaster that you be dismissed immediately and not graduate with your class. You will stay in the infirmary until the ceremony is over. Caleb, this is a truly disappointing finish to a very spotty academic tenure here at NMH, especially in light of your father delivering the commencement address. I can only hope for you and your family that you find a place in the world that feels more comfortable and where you can finally thrive. As with all expelled students, you are considered Persona Non Grata and asked not to return to campus for at least a year." —David Schockett, associate dean, Mount Hermon campus, graduation day, June 16, 1988

By the time I rounded Overtoun and hit the final straightaway, my thighs were seizing, calves turning to jelly. The clock read 32:17. I had a hundred yards to go? More? Less? I hated when things boiled down to seconds. It felt too easy to give in. Then I hit the line with sixteen ticks to spare. I couldn't believe it. A second never seemed so fleeting and so significant at the same time. Everything counts, everything. A guy put a paper voucher in my hand as I staggered around the ropes, lungs heaving, brain filled with hummingbirds. I wiped the tears and sweat from my lashes and looked at my watch.

I'd run my last mile in 6:20. Holy shit. Uncharted territory; I was on another planet. I couldn't wait to tell Chris. The teenager who came in a few seconds behind me wobbled into the ropes and sprayed vomit everywhere, staggering to the grass. I couldn't wait to tell her about that, too.

I started running back for Chris, skirting the oncoming runners, their damp faces etched with pain and effort. I could feel my blood coursing in my veins like electricity, through my heart into my brain. I was buzzing, floating, almost outside myself. I felt warm and dry, despite the cold, damp drizzle. I was in my own bubble, but synced up with everything around me, in tune. I could smell the turf, the turning leaves in the air, even the rain. Everything fit like puzzle pieces, my body floating on a cloud of sighs. A feeling I chased relentlessly here, trying to absorb the world with altered eyes, ripping myself out of my every day.

For decades, the runner's high was believed to be a myth, but a few years earlier, researchers in Germany had proven the science. Through PET scans and chemicals that revealed brain activity, they showed that running boosts levels of serotonin in the brain and triggers the release of endorphins and enkephalins (the brain's version of opium) and endocannabinoids (the brain's version of marijuana). The positive feelings were shown to last up to twelve hours and, in some cases, mitigated mild depression. It's not like getting fucked-up, but rather a palpable infusion of well-being, of emotional connection. Another study showed that in outdoor running, especially trail running, like we were doing today, negative ions—invisible air molecules released by trees that are known to increase oxygen flow to the brain—alleviate the seasonal blues as effectively as Prozac or Zoloft.

I checked the front door of Overtoun on the way for a

quick look at the old butt room, or whatever it became; NMH was now smoke-free. But you needed to punch in a code to open the door now. I jogged up the hill toward the chapel, my breath returning to normal. I spied Chris near London dormitory. She lit up when I told her I'd won a pie, but I could tell she wanted to stay focused and not chat. I wanted to let her know she had three minutes to make her time, but I didn't want to pressure. I drafted in front of her and then she veered around me and poured it on, passing a few younger girls. I trailed behind, watching her arms cut through the mist. There was a purity to her stride, the sway of her hips, and I fell in love again. I could see everything at once. I was drunk on clarity.

Maybe I chose to get myself kicked out of here. I'd admired the kids who got booted—a handful every year, with one usually around graduation season. I loved the tears and indignation of the student's friends, the way they talked about him or her, now a rebel if there ever was one. It was like a funeral while you were still alive. Most of my classmates remember me as the kid expelled on graduation day with his father the keynote speaker. At the time, and for years after, that's how I liked it: personality by reputation, seeping like vapor through the keyhole. But sometimes stories define you longer than you intend and can catch you by surprise. Almost two decades later, after a Vermont Public Radio commentary event where I was reading, the father of an old NMH friend approached me and asked if I was the same Caleb Daniloff who got booted out on commencement day. He clapped me on the back and said, "Man, that was the best graduation story ever." I couldn't argue. I bet no one else has sat in an infirmary with the school nurse, an expelled student, listening to his father give the keynote address about roads less traveled over the PA system. It was a club so exclusive, I was the only member.

More than one psychiatrist has suggested I was getting back at Pa. I didn't reject that theory out of hand. He wasn't the kind of dad who played catch after dinner or took me to ball games. I didn't often see him in my Little League stands. His hobbies were bread baking and beekeeping. He walked around on the weekends in a pith helmet and veil, carrying a silver smoker, trailing plumes all over, pulling writhing gold panels from a three-tiered hive. My arms sometimes ended up pierced with bee stingers, stomachs still attached, that Pa had to tweeze out. I was always struck that these insects would sacrifice themselves like this. Then I'd watch him drip honey over thick, warm pieces of bread.

I thought I should want to be like Pa, but I still didn't know who he was. I was embarrassed that he wanted to be called "Pa" and tried not to call out to him in public. He bathed in my bath water to save money, an intimate act, but never spoke to me about my bed-wetting or his own panic and depression. In Moscow, it wasn't surprising that my closest Russian friends were ones without fathers, too, lost to alcoholism, jail, or other women.

I'd been at NMH three years on my own and that Pa was to be the honored guest, to be robed and capped like me, to swoop in and deem me and my classmates complete, was anathema to some part of me that had grown cold. Perhaps I made Pa a helpless player in a narrative I myself had little control of. But I'm not sure I was that sophisticated, even subconsciously. And it always killed me a little when I thought about him in the woods crying after getting the excruciatingly awkward news from Dean Schockett. I can't even picture tears coming from his eyes, his face always so composed, so dry. I didn't like stepping on insects, either.

But maybe it had nothing to do with Pa at all. Maybe I was just petrified to walk across that stage in front of all those people, those eyes; the fear of that motion, both literal and metaphorical, was overwhelming. To hear my name announced would be sickening. What if no one clapped? My time at NMH had been marked by inertia. Hardly anything had been done with deliberation or forethought. Now I was being told it was time to mature. Suddenly, I could hear the clock ticking again, the one that had been paused years earlier when we left for the Soviet Union. A summoning to alignment. I couldn't bear it. I wasn't ready. I was still thirteen years old inside. I thought growing up was something that happened to you, rather than something you could be told to do, let alone something that you could choose to do. I just wasn't ready. I was scared about what it all meant. Cue up the self-sabotage.

So maybe, like Shorter, I'd come back to campus to regain something I let slip away here. I'd returned to take back my story line.

Chris and I staggered into the gym lobby, where runners were checking times and mingling. I turned in my voucher and was given a warm apple pie from a table of dozens of pies. It was beautiful. I loved the weight of it, the bubbled crust, straight out of a Norman Rockwell painting. It smelled like sweet butter burning in a skillet. We decided against the dinner reception and headed for the car. We drove to a lower lot and changed, peeling off our cold, sweaty clothes, sitting half-naked in the car, laughing as a stream of minivans and sedans rolled past us. What if we ran out right now like pantless loonies escaped from the local nut house? I smiled and tugged on my jeans over my cold, reddened thighs. With my shoes and running clothes, I made a nest on the floor of the back seat for the pie, so nothing would happen to it on the ride home, a swaddled infant. We sat for a while in the lot, breath and body steam fogging the windows looking out over the misty Lower Fields. Then Chris started the car and we pulled onto Gill Road, rolling slowly by the school farm toward the state highway that would deliver us to the interstate. I wasn't sure I'd come back to NMH again. I'd gotten what I came for.

32 minutes, 43 seconds (4.3 miles)
Average pace: 7:16-minute mile
57th place out of 209 finishers

New York City

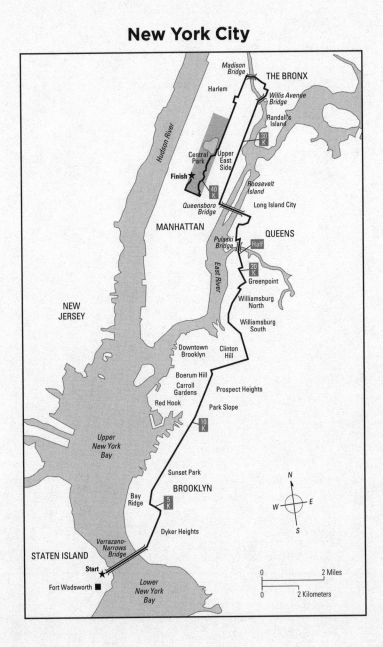

5

40th ING New York City Marathon

Sunday, November 1, 2009

IT WAS EARLY, really early. Even the neon looked tired. Knots of runners crossed Fifth Avenue in midtown Manhattan. Others materialized from darkened side streets and subway staircases. A gathering of zombies in running shoes. Streetlamps showered down pale light and the wind kicked empty grocery bags down the sidewalk. Burly, no-nonsense cops directed traffic, banging on the hoods of idling cabs. A little apocalypse.

Chris pulled onto 43rd Street. I didn't want to leave the warmth of the car. New York City was the rock star of marathons, so massive, so many participants, appropriately epic. At the Fort Wadsworth staging area on Staten Island, almost 1,700 portable toilets were scrubbed and waiting, with 42,000 PowerBars, 90,000 bottles of water, and 45,000 cups of coffee on hand to fill nervous bellies during the long wait before the gun. Once the runners took off, hundreds of volunteers would collect more than 10 tons of discarded sweatsuits, sleeping bags, and blankets, and a fleet of 70 UPS trucks would haul thousands of bags of personal gear to the finish line in Central Park. Along the course, 62,370 gallons of water and 32,040 gallons of Gatorade would be served in 2.3 million paper cups. More than 130 bands were scheduled to perform on dozens of stages in five boroughs, and runners would be able to check their times on the 106 clocks on the route. Waiting at the finish: 60,000 Mylar heat sheets and 52,000 food bags. There was no way you couldn't feel small, like so many plankton inhaled into the whale's mouth.

I breathed out a long sigh and gave Chris a kiss and at the last second grabbed an old dog blanket from the back. It was cold and drizzling and almost five hours until Sinatra's "New York, New York" would cue up the race. Volunteers in orange jackets held megaphones to their mouths like snouts and buzzed instructions. The lines to the idling buses were orderly and swift, and within fifteen minutes we were rolling down the wet, dusky streets toward Staten Island. I'd lived in Manhattan a couple of years in the late nineties as a grad student but was having trouble recognizing the landscape. I hadn't spent much time on the east side of the island and had just been through Brooklyn and Queens once. I had no idea what Staten Island looked like. With the unfamiliar bridge spans and highway traffic lights searing the dark, I might as well have been blindfolded in the back of a panel van.

It was hard to believe this was my fourth 26.2-miler in seven months. I tried to count all the miles I'd run since I'd begun training for my first marathon almost a year earlier—more than a thousand maybe? Almost halfway across the country. Then I scratched out a rough tally of all my miles

ever and came up with a ribbon that stretched a third of the
way around the globe. But what about that terrain that had
no form? Can you tally sobriety? That route had no course
clocks or spectators clanging cowbells or coeds screaming
your name. More often than not, I spaced my anniversary,
and I could never decide if that was a good or bad thing. At
times, it felt like my drunken days belonged to someone else,
to a cousin or an old best friend. Over the past seven years
in running shoes, I'd sweated off much of my past, whittling
myself to my essence, paring knives for hands. How much of
that fat still clung to me?

My grad school classmates might be surprised that the
quiet kid who reeked of bacon grease and tobacco smoke and
dashed out of workshops after an hour to suck down a quick
two in front of Dodge Hall was chatting amiably with someone he'd never met before, about to run for four hours with
tens of thousands of strangers in front of millions of TV viewers. While I managed to be perfectly blunt on the page, I was
terrified of letting anyone in. I hadn't yet learned the difference between honesty and vulnerability. I turned down most
party invitations, skipped class luncheons. I spent more time
with my peers' papers than with them. But they were aspir-
ing writers, too. They understood contradictions and self-
inflicted damage and wanting to live alone in a city of millions. At least that's what I told myself when I locked the door
of my three-hundred-square-foot apartment and got rock-
star fucked-up by myself, porn tapes waiting patiently by the
VCR like groupies. I tried to keep the mess contained within
the pale yellow walls. I stopped going to bars almost entirely.
If no one saw me, then I was OK. Like a kid hiding behind
his fingers.

My seatmate on the bus was a tall Englishman named

James, who was running for a British bone marrow charity. He had shorn, peppery hair and glasses and seemed very awake. He had lived in Brooklyn for eight years and worked for a software company across from the Twin Towers. I commented that the mass transport of all these runners felt like an evacuation for a natural disaster. James said it reminded him of September 11. He'd been at work that day. We didn't say anything for a couple of minutes. I felt self-conscious in my shiny running gear. A former grad school acquaintance who ran the 2001 New York City Marathon described that race, less than two months after the attacks, as a balm for the grief that had draped the city.

"Those were freaky days and I was suffering some anxiety and stress," Oliver said. "I think everyone in New York was probably a little traumatized. There was this one moment at the start where they packed all the tens of thousands of runners onto the starting line and there were helicopters and ambulances and [New York City Mayor] Rudy Giuliani was there. It was intense. Lots of things were being canceled and the city felt shut down. So it felt like a big, defiant party. I don't know if that's how everyone else felt, but it was pretty emotional for me. The cool thing I remember is that there were what felt like millions of people cheering. The route was packed for twenty-six miles with cheering crowds. That was a huge collective act of defiance."

Running as an antidote for trauma, an act of rebellion; I loved it. I looked from the bus window, at the headlights crowding a bridge, the jumbled geometry of the city blocks, and caught a glimpse of my silhouette smeared on the glass.

I wake in stages, as if from an accident. First, taste: the back of my tongue, my sinuses, acrid with cleaning powder residue. A moment later, the smell of burned bacon and stale sweat in the sweater balled beneath my head. Then, exquisite pressure like bamboo splints wedged into cracks in my skull, my brain sparking inside as if stuck with a fork and spinning in a microwave. I force open my eyelids, wincing: blond window shades lit up by fierce sunlight, workshop papers scattered on the floor, baseboards stuck with dust, the phone off its cradle. One leg of my jeans is twisted and pushed up to my knee, no shirt, a rumpled comforter in the corner of the futon couch. It takes several moments for my body and eyeballs to align, and when they do, there's sickening guilt, seeping like vapor through the floorboards looking for memories to attach to. I pray to God there's beer left.

I hang my head over the side of the couch, a thick waterfall of blood behind my ears. I finger one of the stapled papers from under the coffee table and drag it toward me, turning the pages in horror. I've filled every bit of white space with comments and suggestions, even between the sentences. My heart starts thrashing, a bat trapped in a pillowcase. I feel like I've woken covered in tattoos. It's not as if I can white this out. I pull another of my grad school classmates' essays from the spilled stack. Same thing. Front and back. Diary-length scrawlings, with arrows, jumps, and circled words like a basketball playbook. Jesus Christ, I hardly know these people, even after a year. What will they fuckin' think? I'm already gripped by fight-or-flight any time conversation comes near my end of the workshop table, my heart buckling whenever I think about the thesis readings seven months away. I start coughing and stagger to the toilet to throw up. Nothing comes out.

Catching my breath, I sit on the bowl and stare at the grout between the tiles, which reminds me of the Soviet toothpaste that used to line my toothbrush. On the paint-chipped shelf by the tub, I spy the opaque orange prescription bottle. Thank God, Inderal. One or two twenty-milligram pills as needed, the university psychiatrist said. It's prescribed to still the jitters of musicians

and actors, dry their sweaty palms. I shake out five of the bit-
ter blue pills. Right now, just being awake gives me stage fright.
Then the image elbows its way onto my eyeballs, the one that has
followed me the last few years: It's me with my head on a train
rail, my own boot on my skull to keep it in place, while I hold a
shotgun to my temple. Never mind that I'd have to be a hell of
a contortionist or an R. Crumb drawing. This regular hangover
visitor brings measures of terror and comfort. Comfortable ter-
ror? I rub my eyes, pushing them back into my skull. Then I hear
the workmen in the courtyard below. The bat begins thrashing,
sucking the air from my gut. It's fuckin' Tuesday. This all started
three days ago with a six-pack after class on Friday. My neck
seizes and starts to tremble. I make my way to the sink for wa-
ter. I turn the star-shaped handle and lower my face, gripping the
faucet like a tree root growing from the side of a cliff.

Some twenty minutes later, our bus pulled alongside a row
of others near the tollbooths outside Fort Wadsworth, a former military installation on Staten Island. We made our way
to the Athletes' Village within the fort, which was split into
colored zones—blue, green, and orange. A six-language loop
of instructions droned. Some of the faces looked mashed with
sleep; other folks were napping under trees. At first glance, it
looked like a POW camp.

The first New York City Marathon, forty years ago, was a
modest affair, a looping course through Central Park. Of the
127 runners who paid the $1 entry fee, 55 crossed the finish
line, fewer than the number of people on my bus, one of hundreds making continuous trips between Manhattan and Staten
Island. In 1976, when New York Road Runners' founder and
president, Fred Lebow, redrew the course through all five New
York boroughs to mark the nation's bicentennial, 2,090 runners lined up, including two-time Olympic marathon medal-

ist Frank Shorter, along with reporters and television cameras. More than 9,000 people ran in the 1978 race when Norwegian Grete Waitz set a women's marathon world record, finishing in 2:32:30. Waitz would win the race eight more times. Today, with some 43,000 runners pinning on bib numbers (more than the population of Burlington, Vermont), it's the largest marathon in the world, broadcast live to more than 315 million viewers around the globe.

The New York City Marathon's signature is, if anything, the bridges, the city's connective tissue. The race starts on Staten Island at the Verrazano-Narrows Bridge, with runners streaming across both sides of the upper deck and the westbound side of the lower level. After descending from the 2.5-mile-long span, the course winds through Brooklyn for the next 10 miles, surging through a number of neighborhoods, including Bay Ridge, Sunset Park, Park Slope, Bedford-Stuyvesant, Williamsburg, and Greenpoint.

The Pulaski Bridge, connecting Brooklyn to Queens, marks the halfway point. About 2.5 miles later, runners cross the East River on the 1.4-mile-long Queensboro Bridge, which dumps them into Manhattan. The race then heads north on First Avenue, crossing into the Bronx on the Willis Avenue Bridge. A mile in the northernmost borough and it's back to Manhattan via the Madison Avenue Bridge. Then south through Harlem, down Fifth Avenue, and into Central Park. At the southern end of the park, the race cuts across Central Park South until it reaches Columbus Circle, where runners reenter the park and less than a mile later greedily, thankfully, joyously, painfully cross the finish line outside Tavern on the Green.

The sun had now risen but was being held in check by thick clouds. A light drizzle was spotting my glasses. I guessed the temperature to be between 47 and 50 degrees. A large communal tent had been erected on a patch of grass, but every square inch was taken, with legs and feet sticking out. So James and I sat on a curb abutting a strip of grass. We shared my dog blanket, discussed training and shoes; the sentences almost felt natural on my tongue. Some weekends when I'd lived here, I'd barely opened my mouth except to let the beer in and the smoke out. Classrooms continued to riddle me with panic and doom; even formulating basic sentences in my mouth was a monumental task when eyes were upon me. But I stuck with the terror I knew best. A university shrink wrote me prescriptions for pills that muted the jackhammers in my heart and kept my mind from overrunning me. Sometimes, I believed those tablets were the only barrier between me and a public heart attack.

James seemed impressed when I told him I'd studied in Columbia's MFA writing program. A lot of people were. I know I once was. I had the creamy vanilla paper with embossed black script framed. But for years after, my MFA stood only as a reminder of what wasn't. I'd mistaken it for a literary achievement. I'd figured I'd have a book deal upon graduation, that agents were hiding behind every tree on campus, faces pressed up against our workshop windows. I kept a running draft of the acknowledgments page for my yet-to-be-published, groundbreaking memoir. Publication was going to change everything, justify everything, be everything.

The difference between veteran runners and newbies was obvious. The newbies were cold. Those who'd run before had swaddled themselves in sleeping bags, Mylar finishers' capes, soft hooded hazmat-style suits, and long ratty down coats. I had on three layers plus my North Face windbreaker, and a hat and gloves, but could still feel the chill. As the drizzle turned to light rain, James took off his shoes and put them in plastic bags. He'd also brought a knapsack to avoid fighting the crowds at the end to retrieve his gear. Tall man traveling light. I liked it. We snapped pictures of each other against the looming Verrazano-Narrows Bridge, with its huge towers and plummeting support cables, a reminder of the vast task at hand, and e-mailed them to our wives who were waiting back in Manhattan.

At about 8:30 a.m., the ferries arrived from Manhattan, depositing thousands more runners at Fort Wadsworth, eating up already scant sitting space. I packed up my jacket and windpants in the official plastic bag and laced up my Brookses. I wished James a good race. We parted ways and I headed for the green camp and the UPS truck that would bring my gear to Central Park. Runners were tossing their unwanted sweatpants and jackets onto mounds of clothing, like unlit bonfires, that would be donated to charity. A couple of white hazmat suits lay flattened on the ground as if the occupants had been vaporized. I tossed my blanket and sweatshirt, too. An offering to the gods.

I wipe my mouth on a towel on the back of the bathroom door and shuffle to the kitchen. The coffee machine clock reads 1:47 p.m., the red digits glowering at me. The pot is full and cold. I grab the Camel Lights from the counter. Empty. I drop the pack to the floor and scan for another. I notice there are no bottles anywhere. Not a one. I'm confused. Then it hits me. The homeless guy. I can't remember his name. He was black, skinny. I met him near the church on Broadway and brought him back to my apartment last night, promising him my empties, about seven grocery bags. We sat on the couch and listened to some Stones. I'd offered him a line, but he turned it down. I have no memory of what we talked about or of him leaving.

There had been other visitors, too. The Billys. Small Central American coke couriers. Three or four times they were summoned. The last Billy had asked me if I was all right as he dug the fold from his messenger bag. My heart was racing, eyes bugging, face vibrating. I'm fine, I'm fine, I'm fine, here's your money. Then I'll rip up your pager number, for real this time.

Bits and pieces of the last three days start to materialize like prints in a developer's tray. I can see Jack, my old high school pal, pushing me out the front door of his place on East 75th. Angry? Or laughing? I remember a long cab ride. Shooting pool with some classmates at a bar on Amsterdam. Please tell me I didn't burst into my Axl Rose snake dance. Why do I still do that? I'm almost twenty-nine years old, for chrissakes. I remember critiquing papers, one after the other. Then walking down to 108th Street and up Amsterdam to stare at the Cathedral of St. John the Divine. I hit the hot bar at the West End Market and walked to Riverside Park. It was like trying to follow a perverted "Family Circus" cartoon, where Tommy, sent to borrow a cup of sugar from the neighbors, ends up zigzagging, backtracking, creek-hopping, tree-climbing all over the neighborhood. I made a lot of phone calls. To Russia, to high school friends I hadn't spoken to in years, to Chris. Shit.

I joined the herd of green bibs headed for the start, shuffling through a field that was more mud than grass. A light mist hung in the air. Five minutes later, I was crushed when I figured out that we green bibs would be starting on the lower deck. Everyone else got the open-air veranda, a vault of blue sky overhead, the Manhattan skyline gleaming ahead like a row of liquor bottles behind a bar. I wanted to be a pixel in that iconic shot on the upper span, the mass of humanity covering every square inch of that mile-long bridge that would be on the front page of the *New York Times* the next day, a small, swarming city moving across a marvel of weight-bearing engineering. We were in the root cellar.

But I didn't have long to moan. Without warning, the folks in front of me started shuffling, then jogging, and we picked our way onto the covered deck. Yelps and hoots and whistling began ricocheting off the ceiling and supports like madness. Four hours of pent-up excitement and nerves bounced around our heads, so loud my instinct was to duck. Runners stopped to snap pics of the New Jersey shoreline across the Lower Harbor, others to piss into the Narrows that separates the upper and lower bays. Streams arced down from the upper deck, too. It felt festive until something wet hit my thigh. I told myself it was a rainy leak above and steered my body toward the middle of the pack.

By the time we spilled into the light of Brooklyn at 92nd Street, a poisonous thought seeped into my brain: *I'll be glad when this is over.* Shit. I'd ripped myself out of the moment and we'd hardly started. I wanted the experience, but as memory. I wanted to fast-forward. This wasn't good. I tried thinking of things that took four hours—an overtime football game, *Gone With the Wind*, the bus trip from Boston to New York—but that didn't help. They all seemed interminable. Why was I suddenly disengaged? I had a list of reasons, of course: the long, cold, rainy wait; burnout from three previous marathons; I hadn't taken a dump in thirty-six hours. My mind refused to juice my NMH triumph two weeks earlier. That was only 4.3 miles through an idyllic wooded prep school campus, not a marathon through a concrete jungle.

But I reminded myself that every run, especially a marathon, was about conquering doubt. On my daily runs, I often started off nervous, never quite sure how the miles would sugar out. On mornings that I felt heavy and tired, blow-torches blasting my knees, I ran sub-8:30s. But on some days when I felt rested and fortified, I'd have to fight through the bacon feeling. Despite the sameness of running's mechanics, the swirl inside could be as touchy as mountain weather. I slapped myself. *Run tall. Stay in the moment; forget your mind. Have fun.* It was that last part I worried about.

My plan was to clock 8:40-minute miles the first half of the race. I'd need to average 9:09 to break four hours. In Moscow, I'd faded in the second half so I figured I'd bank as much time as possible—almost thirty seconds for every mile; that gave me an extra 6.5 minutes to play with—to make up for any late-stage plummets. My first mile was 9:34, but I chalked that up to the tightness of the crowd. My second mile near Dyker Beach Park was 8:55. I was starting to feel a little better. I held my next three miles at 8:50.

A chunky necklace of spectators lined Fourth Avenue. Women in bathrobes sat on the stoops of brownstones. Girls in sleeping bags were curled in windows and on fire escapes. Barbecue smoke wafted through the air carrying the smell of sausage and burgers and grilled peppers. Cops on motorcycles swept the streets, keeping overzealous fans off the course. The clusters of runners still hadn't broken up. It was like snaking down a subway platform at rush hour. Space in this version of New York, fittingly, was at a premium, but picking my way into the open spots burned valuable energy. I was sure the precious glycogen stores in my legs and liver were depleting with every weave and side step. Around my ears, clouds of German and Italian and Spanish; some ran with their national flags tucked in their singlets. Other runners were chatting on cell phones. Some snapped pics while jogging backward. They looked so easy, like they had the secret formula.

Every street corner seemed to pump out live music—rap, reggae, brass, Guns n' Roses covers. I slapped a lot of hands, kids, adults. Each one felt intimate. I thought about their lives, how they would spend the rest of the day, the week, why they were drawn to this early-morning spectacle. I began to feel connected, the cables plugging in.

In Park Slope, I came across a short, thick Spanish man with JESUS Magic-Markered on his shirt. He was running on two Cheetah blades, his legs swinging out and back in like scythes over a field of asphalt. Spectators called out his name, but Jesus stayed focused. Beyond him a wheelchair-bound runner was pushing himself backward with his feet. Then something I hadn't seen before: a runner with no legs at all, just an upper body. He used a single centered Cheetah blade and arm-brace crutches to help push himself forward. I felt guilty passing him and touched my fingers to my forehead in salute. I tattooed his image on my mind for later in the race, to help me through that often wrenching no man's land of the last six miles.

I caught up to a foursome—two men, two women—with bib numbers that read 3:50. One of the male runners was holding a balloon stamped with the same digits. Official pacers, cool. All I'd have to do is stay with these guys and I'd still have a 9:59-minute cushion. I dashed ahead to the fluid station and grabbed a Gatorade and chased it with water. Then swam back into the mix. I felt like I was keeping an even stride, so I was surprised when mile 8 clocked in at 9:34. I felt my breathing scraping a little deeper, my legs tightening. Then over the next few miles, my watch coughed up a couple

of real loogies: a 9:47 at mile 9, then a 9:58 at mile 10. Why
was I dragging? Were my shoes too tight? My toenails too
long? Hold on, where did those pacers go? I pressed on, scanning the mass of bobbing heads for that magical orange and
blue balloon. Nothing, I was adrift at sea. *Follow your breath.
Run tall. Have fun.*

A half-mile later, a calm washed over me, and my music seemed crisper, louder in my ears. Everything sharpened.
I looked around and saw hardly any spectators as if they'd
been erased. Bearded men in long coats and fedoras strolled
the sidewalk; one leaned against a lamppost and watched us
streak by. I'd reached Williamsburg, the large enclave of Ha-
sidic Jews that was already becoming a hipster village. The
women wore dark, bobbed wigs and pea coats and pushed
strollers. The boys, in white shirts and earlocks, looked like
members of the same gang. I didn't see any girls. The hipsters
must have still been in bed.

The buses, Fort Wadsworth, the throngs of runners, all felt
left behind, blown away like dandelion seeds. I was alone running. The interlude in this bubbling, wild race was welcome,
a reminder of the blank slate of running, that opportunity to
redefine myself. I turned off my iPod out of respect for the silence. In my ears, the sound of my heaving lungs like crash-
ing waves. I could hear myself in the act of living, reminded
of the greed of my own existence. I pictured myself, alone in
a frame, in full color, the shadows of other runners knifing up
behind me, but never nicking me. Not a care in the world. I
owned this road, this unfamiliar neighborhood, this great city.

One of my greatest joys was setting off in new cities and
places I was lucky enough to visit—the Reykjavík waterfront,
the Los Angeles canyons, the coastline of Isla Mujeres—tak-

ing me past nooks and crannies I might not normally see, re-
newing in me a sense of exploration, of wonder. I was not
only logging new geographic turf, but claiming psychic terrain as well. When I was drinking, the only thing I wondered
about was where my next drink was coming from. Now, I described new run routes with the same zeal I used to recount
nights of debauchery.

A couple of miles later, I spied the red arches of the Pulaski Bridge, a 2,800-foot drawbridge-style span that normally
carries six lanes of traffic and train rails between Brooklyn and
Queens. Running parallel to the East River, the Manhattan
skyline glittered to the west. The halfway mark. I was closing
in on a 1:58 split, so a sub-four was still within striking distance. A man in a ratty leather bomber jacket and knit cap
held a sign that read GET OUT OF BROKLYN! [*sic*].

Then I saw a tall runner holding a giant balloon on a
stick, which was markered with 3:59:59. I lit up. He was surrounded by a throng of sweaty runners. My people, following
a Pied Piper pacer. I watched that sweet sphere bouncing like
the dot over the song lyrics in an elementary school movie. I
was grateful that this guy took it upon himself to help the numerous runners pouring across the bridge with the same magical number tattooed on the backs of their eyelids, on their
hearts. It was touching, the silent camaraderie. I wanted to
thank him for ushering us toward that golden barrier. He understood the vast space contained within a single second. And
I loved him for it.

I pushed harder for a few minutes, reaching the Pulaski's
midsection. *Wait a minute, why was my guy speeding up? Where
was he going?* I tried to ratchet up my pace, but my legs had
sprouted barbells. My body would not obey. All I could do

was watch the balloon grow smaller until it disappeared. Then the 4:00 pacers passed me. They might as well have whacked me in the face with a sockful of manure on their way. I could have cried. I let the feeling pass. I used to get mad at things I couldn't change. I'd curse out loud at the sun for strobe-lighting its rays through the tree branches, a gusty wind for pushing me around like a bully. But the more I ran, the more I learned to accept things beyond my control. *OK, breathe.* I recalibrated. If I held my current pace, maybe I could still post a 4:09:59. Still a respectable satin sash to strut around in.

As I approached a water station near McCarren Park, a tall man in a daisy-covered singlet threaded his way through the thirsty mob and snatched up a cup. It was James. I couldn't believe it. At first I was chuffed that I'd kept up with such a long-legged runner. The shiny look on his face suggested he was having a grand time, his arms swinging in tune with the steel drums, gospel singers, and high school jazz band along the sidewalk. I was amazed at the odds. I wanted to dash up and slap him on the back, chat a bit. But he didn't linger and rolled back into the fray, his backpack swaying. Plus, I knew I didn't have the energy to catch him, let alone sustain chatter. So I let that moment go, glad to have met him, a reminder that paths that cross will often cross again. I found myself smiling for the first time in several miles. There was an order to this sweaty, mobile universe and I was a part of it.

I tried to stick to the flatter middle part of the road since the sides slant to sluice away rainwater. But the bodies were still clumped pretty well around the double lines, and other ankle-minded runners had the same idea. Over the next few miles, even the fluid stops became stations of treachery. Littered cups had been flattened into a waxy pulp with some runners skidding through like skaters. At another station, banana peels littered the ground. I pictured a cartoonish scene of hundreds of tired legs giving out from under, a spinning blur of asses and teakettles.

Next milestone: the Queensboro Bridge, or 59th Street Bridge, a 1.1-mile span draped in caged ironwork that soars 130 feet over the East River, across Roosevelt Island, and spills into Manhattan. Again, it was lower-deck action all the way. The party chanting was in full effect, rubbery echoes bouncing everywhere like dodge balls. It was a long slog that felt more like one of those treadmill stress tests. I picked up my pace on the descent to chip away at lost time. But I wasn't really in control anymore. I was a torso on top of cement legs in roller skates, imprisoned by its own momentum. I just hoped I could make the turn and not run straight into that tree. You know, the one that grows in Manhattan.

But the bridge was only a preview. Almost four miles of First Avenue lay ahead, much of it an incline. A fair number of people were walking. I was clocking 10:50-minute miles and starting to become a hater. I'd hit the wall at mile 14, and for tradition's sake, the wall hit me again at mile 18, knocking my pace into 11-minute territory. I shuddered at what lay in wait at mile 23. I kicked myself for my fast start, my shitty bank strategy. *Whatever you do, don't stop.* The mortal sin. I was afraid I wouldn't be able to start again, afraid of the pain of that very first step.

My inserts began digging into my toes. I still used the confetti-colored microwave orthotics from six years ago along with Dr. Scholl's three-quarter inserts. Together they were like slices of layer cake beneath my socks. The arch contours had long ago flattened out but provided me mental comfort. Ev-

ery time I switched them out, my legs hurt and I felt unbal-
anced. I'd become superstitious about running, I realized. I
always ran back to the exact place I started from. I had to
touch something at my turnaround point—the street pole,
bridge railing. No stopping. I'd developed rituals and routines
and negotiation tactics, same as when I smoked and drank.
*Just another hour and you can crack that beer. Just another mile
and you can call it a run.* The deviousness I once honed to
fool myself, the one-time comfort I had with public spectacle,
the perseverance I applied to keeping the buzz going, could
still be harnessed. *Just make it to that next lamppost, to the next
block, to the next fluid station. Keep it flowing. Who cares what
you look like.*

Heading into the Bronx at the Willis Avenue Bridge, rubber carpeting had been laid down over the gridwork surface,
the first bit of soft relief. A man was handing out roses while a
lone bagpiper wailed. I was running through a wake in someone's living room. I wondered where we were in relation to
Grand Concourse, the borough's main thoroughfare. But I
probably wouldn't recognize it. Jack and I spent some wee
hours there as he tried to track down his dealer, who'd taken
his cell phone for a drug debt. I'd loaned him the money to
get it back. Back in the cab, Jack got to business, nose buried
in a paper fold, and we went back to his place. He put down a
line for me. Part of me was touched. I'd never done heroin but
had secretly aspired to. I'd met few lines I didn't want to cross,
or put up my nose, and I was pretty lit. But there was pause, a
catch in the throat of my mind. I thought about Chris. Even
if she never found out, I knew it would change everything,
create a distance between us, if only in my heart. Looking at
that brown line, I knew I'd be leaving behind even the famil-

iar booze-soaked cavern I'd carved out. I'd feel tainted around Shea. Because I would know. Shea represented pure, unconditional love. Love without a price, for its own sake. Maybe I wasn't ready for this type of feeling and relationship, scared of it, scared of having to quit drinking, scared Chris wouldn't like my sober self. Was I just hiding out in grad school to avoid the inevitable? I'd already been flirting with the idea of a PhD. I knew Chris and Shea were my last buoys before open, endless sea.

As I stood there, the cut straw inches from my fingertips, I knew I was on the verge of sabotage. I felt around inside for that pebble of consciousness, that eye in the mud that remained sober. To consciously disobey that tiny voice, which was trying harder than ever to be heard, would mean I could never trust myself, my black box lost, never to signal again. Jack had gone into another room as if he didn't want to pressure me. I wasn't as strong as Jack when it came to drugs. I always took things both too seriously and without much thought. I stepped back from the table.

In the fridge, I find three beers. A collection of shining suns. I crack one and light a cigarette. A cool warmth blossoms in my chest and spreads to my brain, melting the fog, dulling the aches. I put the phone in its cradle and look at the answering machine. The display blinks six. I press Play. My mom twice. Nate. Two hang-ups. A wrong number. But no Chris. I remember talking to her—last night?—not letting her off the phone to put Shea to bed. "I just want to read you some new paragraphs real quick," I said, trying not to slur or sniff. Yes, OK, that was Sunday night. I remember because I'm pretty sure that's when Jack called and I said I'd call her right back. Did I? Fuck.

I drain my beer and dial Chris's number, trying to figure

out what sort of tone to muster—neutral? Excited? The phone rings several times. I hope Shea answers. Maybe if I can make her laugh and giggle and ask when I'm coming up again, Chris will soften by the time she gets on. Chris picks up and she's cold.

"You remember what we talked about last night?"

My mind is blank. I know by her tone it's bad.

"Sure," I say, eyes darting for the Inderal bottle.

I pull open the fridge and slip out another pounder, holding it at arm's length away from the receiver while I gingerly twist off the cap.

"I said that I didn't think this was going to work anymore."

The bat tears through the pillowcase and hurls itself at the window, even though I've always known this was a possibility. Ruined relationships rattle behind me like cans strung to a bumper. But it's different when it's actually happening. Hyperreal, like every camera angle crammed into the same frame. I'm a Hockney collage. This wasn't the first "serious" talk we'd had about my drinking, but it felt like the last, her words clipped, edged with finality.

"We're leading very different lives," Chris said. "We're too far apart. The phone is not enough. You don't always call when you say you're going to. Even when we talk on the phone, or you come to visit, I'm lonely. You never go to bed until the beer's gone. Shea can't be around all this."

All this. *I feel sick. What if Chris's ex gets wind and makes a stink? He hates Chris now. And me, too: a home wrecker in his eyes. What am I doing? These are people's lives. Chris has already upended her and Shea's life with a divorce. This isn't college anymore. This isn't girls who don't mind making out in the kitchens of house parties or in the corners of bars, whose parents drive Range Rovers and finance ski weekends. I don't say anything. I thought we were destined to be together. But that's what*

I thought about a lot of girls, throwing around marriage talk to some within weeks. The cliché of this pattern makes me nauseous. Here I am, another in a crowd of a million drunks flushing everything down the toilet, afraid to face life without six-packs for hands.

"OK, I'll quit the beer," I mumble, the words like stones in my mouth. "I want this to work. This is important to me."

The word beer *feels less absolute than* drinking. *I wonder if that's a mistake. "The drinking," I add.*

Chris doesn't say anything. I hear Shea in the background. Shea who now spends three days a week with her dad.

"I have to go," Chris says.

I hang up and drain the last Bud in the fridge. "That's it," I tell the coffee table, sucking down the foam from the bottle. I stare at the brown glass, at the red and white label decorated with a Founding Fathers–style script, making a mental note of my last drink, a postcard to tack to the fridge. But nothing stands out about it, looks just like all the others. I switch on the TV, now at the mercy of time. Fretting. Smoking. Watching Oprah. *Watching* Seinfeld. *Waiting for the thirst to set in. For now, there is something approximating relief in my chest, but it's not absolute. In the corners, the cold heat of nervousness, uncertainty.*

I know I have just quit drinking for Chris, to save our relationship. Out of fear. To not be alone. For reasons other than for myself. The AA folks say no romantic relationships the first year. But who am I without a girlfriend? Without booze? To suddenly have neither feels like a murder-suicide. I light a cigarette. Does this mean moving back to Vermont after grad school? Is that it for being a writer? So many unknowns swirl through my mind, and I've just sworn off the one true thing I know. The ledge rapidly disappears above me.

The rest of the Bronx was a blur. My body felt like an

overbroiled holiday roast. I could hardly lift my head to look at the giant monitor stationed on Morris Avenue. If I'd seen the pain and disappointment and exhaustion on my larger-than-life face, I feared total collapse. I wished I'd packed ibuprofen. Maybe I shouldn't have taken four days off. Maybe I should have done some training research. I didn't understand tapering. I was creeping up on twelve-minute miles, almost a brisk walk. Maybe I put too much stock into this race. Wouldn't it be perfect if I broke four hours here, a symbolic nod that I'd come out the other side?

Now, it was looking more like New York was the place where I not only hit bottom—but the wall, too, multiple times. But as I slogged back into Manhattan, I knew that old pebble of consciousness was still there. I recognized that, over the years, it had grown into a stone I carried in my pocket like a talisman and later a rock I could cling to. And I knew its potential—a boulder to lean against, and eventually a mountaintop to stand upon. From there, all horizons would be visible, the lush valleys and where the forests still smoldered. From there, the final tally would be conducted. *One foot forward, then the next.*

I trudged past the spiraling Guggenheim and tony apartment buildings with stone-carved balconies. When the hell did a hill go into Fifth Avenue? It just kept climbing. I was Sisyphus and my body the boulder. The music, cowbells, horns, wolf whistles, and reaching hands were overrunning my senses like a mob. The cheering had morphed into a sustained cacophonous roar, rendering my iPod headphones useless. Runner after runner passed me, some cheering back to the crowd and fist-pumping. I almost resented their energy. My insides had gone into fetal position. Was this why I ran marathons, for those last six punishing miles? For the clarity of pain?

When I finally quit drinking, not putting my lips to a Budweiser wasn't as difficult as I'd always feared. It came down to a choice: a future with Chris and Shea or one with the bottle, alone. Sometimes it felt like I simply picked to go right instead of left. I wasn't even sure I'd hit bottom; there was still more to lose. I guess there always is. I muscled through a few days of sweaty sheets, jitters, and wicked thirst. The sluggishness and headaches lasted a little longer. But I still went to class, still bellied up to my keyboard, still critiqued my classmates' stories, eventually even made a few party appearances, seltzer bottle molded to my palm, the carbonated burn trying to fool the beer gland in my throat. In fact, at first, saying, "No thanks, I don't drink," felt empowering, intriguing, special.

The initial period of sobriety is often called "the pink cloud." There's a feeling of acceptance and excitement, a shine. Things look, smell, and taste different. There's a different appetite. Early sobriety feels like a fresh path, overgrown and lush, buzzing with life, what most people had long trod smooth. Those first months, I cleaned my apartment top to bottom. I read. I went to Vermont. I watched movies. I went to readings at the 92nd Street Y. In the evenings, I drank seltzers and smoked cigarettes. I still popped my pills before I left my apartment. Broadway remained unpredictable and filled me with vague dread. I always screwed my earphones in. I even managed to pull it together and read at the annual thesis reading. I was an explorer standing in place, head mildly buzzing, with little idea of the length of the road ahead of me.

They say that alcoholism is the fear of life, and when fear

rules your days, that's all you feel. As a drinker, I mistook the presence of fear for cowardice. I was terrified of conflict, something both amplified and brought on by the bottle. Even after I quit drinking, I stayed scared, the fear ingrained in my fibers. I began to believe this was who I was, born afraid. It took years to realize that courage isn't the absence of fear. It's being scared and acting anyway. No, that's why I ran these marathons. For what they represented: the physical manifestation of going through something, not around it. I was looking at myself with a raging, surging heart, and not turning away. Fear makes me human. Running allows me to act.

Finally at 86th Street, we wheeled into the park, just below the reservoir. The sloping roads, trees, and greenery were a welcome sight. It was refreshing to separate from the traffic lights, street-cleaning signs, and metal trash cans. It injected softness at just the right moment. The bite in my calves started to loosen. Behind the grandiose Metropolitan Museum of Art, leaves were strewn along the asphalt like petals. Then I spied the mile 24 sign. A beautiful sight, like spotting land.

Two years earlier, near this spot, twenty-eight-year-old Ryan Shay collapsed and died during the United States Olympic marathon trials in Central Park, held the day before the 2007 New York City Marathon. An undetected heart condition. A lot of runners think of Shay near East 72nd Street. Even though he was felled in his prime, it didn't seem like a bad way for a marathoner to go, to leave your heart, your final muscle twitch, on this monster course. That's how I might want to go. But then I thought of Shay's family and friends and was pretty sure they didn't think this way. I bit back on the romanticizing—and the complaining.

We streamed out of the park and onto 59th Street, or Central Park South. I don't remember much of this stretch—the long stone wall, the octagonal-paved sidewalk, the faceless crowd boiling over the barriers—my mind simply wasn't accepting any more information. It was too focused on getting me across that finish line and ending the pain. Seeing that mile 26 marker was almost overwhelming. Just the .2 now, a little more than three football fields.

For the 1908 Olympics in London, the marathon was changed from 24.8 miles, the distance from Marathon to Athens. King Edward VII and Queen Alexandra had requested that the race begin at Windsor Castle, 20 miles west of central London, so that the royal family could view the start. The Olympic stadium in London was 26 miles away. Then event organizers added an extra 385 yards around the arena track, so the competitors would finish directly in front of the king and queen's royal viewing box. Damn royals.

Within a couple of minutes I saw the finish mat and giant clock. I managed to raise my arms for the camera as I staggered across. From go-go-go to dead stop, no reveling, no basking, no victory laps. Before I could blink, a goody bag was pushed into my hand. I dug out an apple and bit, the juice running down my chin. I pounded the water and the Gatorade. I followed the sea of roiling foils headed for the UPS trucks to get our gear. I felt like I was walking on peg legs. I was cold, clammy, drained of whoever the hell I was. What more was there to say? I was done, no more races, no more wall, no more staging areas. Four marathons in seven months, plenty for people to be impressed with. That's what I told myself, but somehow I wasn't convinced. I still felt incomplete. I still didn't have my sub-four.

Two days later, I woke up, brushed my teeth, and marked the Marine Corps Marathon's registration date on my calendar. It was time to go home.

4 hours, 22 minutes, 12 seconds
Average pace: 9:50-minute mile
22,904th place out of a world-record 43,660 finishers

Middlebury, Vermont

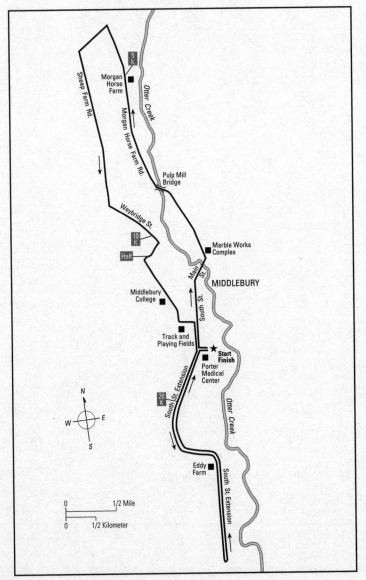

6

2nd Middlebury Maple Run
(Half Marathon)

Sunday, April 25, 2010

A S WE DROVE AROUND, I didn't see much that had changed, as if the place had been vacuum sealed. I'd lived in this small college town twice, as different people—for two years as an alternately angry and shy twenty-something drunken cub reporter and then, after a break for grad school in New York, for seven years as a dry, thirty-some-thing, struggling-writer stepfather. Gliding along Route 7, past the elementary school and county sheriff's office, I was unsettled by how familiar it all felt. Chris and I could have been running Saturday errands—dry cleaner, Hannaford, Ag-way for dog food. We pulled into the Shaw's grocery store parking lot for some Gatorade. Chuck, the dough-faced cart guy, who used to mistakenly call me Kyle, was steering a train of carts through the automatic door. He still sported a thick goatee and oversize green Shaw's shirt, a couple of homemade

tattoos squiggled on his forearms. He made outrageous claims
like he'd lost twenty-five pounds in a week or that he was connected to the Italian mafia. Because I once got into a conversation with him at the gym, Chuck saw me as his workout
mentor, picking my brain about curls and crunches and diets
in the cereal and bread aisles as if he couldn't see the daddy
belly clinging to my front like a koala bear. Even though
I'd been working out for only a year, I felt like I couldn't let
Chuck down and always coughed up some kind of half-baked
fitness advice. I was Kyle the personal trainer.

Afterward, we stopped by the Middlebury Inn to pick up
my number. I'd never stepped inside the iconic Victorian hotel across from the green, where college dignitaries stayed and
the town's cultural bigwigs held silent auctions and charity
parties. The paper always covered them, photo essays showing
the same group of notables — poets, doctors, lawyers, painters, business owners, in tuxedos and fancy gowns, grinning
and dancing the night away. Even cutting through the parking lot on the way back from Video King (where Merchant
Ivory films were shelved in the Foreign section), a Spice Girls
tape in my hand for Shea, I'd felt like I might be asked for my
pass and then be told to keep moving.

The hotel lobby had a dark, woody feel that made me feel
like grabbing a black coffee and writing a poem about a one-
room schoolhouse. An oil painting of Robert Frost hung on
the wall near a clutch of caramel-colored wooden tables and
couches. I stepped up to get my packet, and the guy who
tossed me my T-shirt was none other than my old lawyer.
Benj was one of the race organizers. I had no idea. Mine had
been one of hundreds, if not thousands, of hands Benj had
held through the Addison County Courthouse over the years.
At twenty-six years old, I'd ended up in my own newspaper's

police blotter. Benj looked thinner. Running had apparently changed him, too. Our eyes met and there was a spark of recognition, but what was I going to say? "Hey, Benj, remember me? You got my dope charge dropped fifteen years ago." Then the next runner walked up for his packet and we both moved on.

"*At approximately 2345 hours, I observed a 1994 blue/black Saturn, bearing Vermont registration BGS822, traveling south in excess of the posted speed limit of 50 mph. This speed was confirmed on the radar at 70 mph. I turned around and stopped this vehicle on US Route 7, near the intersection with Exchange Street in the Town of Middlebury and the defendant, Caleb S. Daniloff (DOB 12-01-69), was identified with a valid Vermont driver's license as the operator and sole occupant of the vehicle.*

"*Upon speaking to the defendant, I noticed a moderate smell of intoxicants coming from the vehicle and the defendant's eyes appeared to be bloodshot and watery. I noticed there were six (6) empty Budweiser beer bottles on the floor in front of the front passenger's seat. I asked the defendant if he had been drinking and he stated he had been drinking at a parking lot in Vergennes, and he had approximately four (4) beers the last one being approximately 1/2 hour ago.*

"*When the defendant exited the vehicle I noticed what appeared to be a 'pot' pipe sitting on the center console next to a small wooden box. The defendant admitted the pipe was his and he had smoked marijuana earlier in the evening. I removed the pipe and box from the vehicle and upon opening the box, I found another 'pot' pipe and a small amount of green plant-like material. Located under the front passenger seat I also found a clear plastic bag of dried green plant-like material.*

"*The defendant gave an alco sensor at approx 0005 hours the result being 0.12 BAC%. The defendant failed the alphabet test*

and the one leg stand test. I then asked the defendant to per-
form another dexterity test and he stated that he did not need to
as he knew he was over the limit. The defendant was taken into
custody for Driving While Intoxicated."—Senior Vermont State
Trooper Thomas L. Hodsen, April 12, 1996

The next morning, race day, I ate one crumpet with pea-
nut butter, a banana, and three strips of beef jerky. I washed it
all down with Gatorade and water mixed with vitamin C energy powder. Exactly what I'd knocked back before a training
run a week earlier where I'd clocked 13.1 miles in 1 hour and
47 minutes, shaving 6 minutes off the Hyannis Half Marathon I'd run while training for Boston; my last 2 miles, in
fact, I'd reeled in 7:15s, a scorching, superhero pace for me. I
was brimming after that one, ready to toss aside school buses
and shake the sidewalk out like a dusty rug. To post a 1:47 on
an official race results page would juice the feeling tenfold.
Like a chemist in the lab, I was doing my best to re-create the
sequence and conditions, even wearing the same shirt, shorts,
and socks. I was always seeking that perfect run, still chasing
the perfect high.

I pulled into Porter Hospital at the end of South Street,
just beyond the starting area. I changed out of my pants and
hoodie and pinned on my number. Other bibbed competitors
were warming up on the dirt road beyond the parking lot,
my old run, the part of Vermont I missed most since moving to Boston. The twisting gravel lane lined by cornfields,
horse paddocks, and swampland made up the last five miles
of the race. I'd logged countless hours on this hard-packed
dirt, mostly in the predawn. It was here, starting out in the
black of unbroken night, that I finally began mastering the
transition from darkness into light, literally practicing it hundreds of times over, all against a soundtrack of insect chirp-

ings, birds rustling in bushes, the cracking of frozen branches. Sometimes, I swear I could hear the fog breathing. In the sweaty, heaving darkness, running and dreaming became one—curled leaves were dead birds, trains pushed balls of light through the trees, mailboxes were giant roosters. Guided only by starlight, I had no form, the world underfoot shifting and malleable. From somewhere in that surreal mix, I began to materialize, returning to the blacktop a different person from when I woke up.

Looking at the lean legs tramping across the divots and washboard ruts, bib numbers snapping against their chests, ponytails whipping the air, I saw only interlopers in the church pews. South Street Extension was mine, sacred, a secret fishing hole. It was one of the few places where I'd fit in this small town. I wasn't a local. I wasn't with the college. I wasn't a hippie alpaca farmer. I didn't move here to raise my kids. I wasn't the church-going type. I'd moved back to be with Chris and Shea, to start a new life, away from distractions and temptations. The pink cloud had drifted on by the time I'd finished classes at Columbia, revealing below giant craters, a dry, dusty moon surface, and I'd returned to an old Zip Code with insecurity and confusion baked deep into my bones, seeking exile, as well as another chance.

I took a staff job at a daily paper an hour south of Middlebury. Clutching a pen and reporter's notebook, I had no problem talking to strangers, but among my new editors and coworkers, I fell apart inside, panicky and unsure. *I used to be fun and witty and interesting,* I wanted to say. *You want references?* I covered select board meetings I didn't understand and house fires where the firemen asked me to help carry water canisters to the scene. I wrote crazy ledes and peppered my stories with colorful but irrelevant details. I still believed my

MFA had stamped Future Literary Star on my forehead like a Harry Potter lightning bolt. I was convinced the appropriately disjointed and ill-formed memoir I was hoping to shop soon would change my life, make me rich, and retroactively convert my awkwardness and standoffishness into delightful eccentricity. I made sure copies of the *New Yorker, Harper's,* and the *New York Review of Books* were prominently scattered about the apartment, a small two-bedroom in a converted funeral home. A writer lives here, goddamn it. Not a jittery, ten-month-sober, lily-hearted reporter trolling the Diamond Run Mall with that week's Man-on-the-Street question and a point-and-shoot. *Excuse me, sir. What's the worst Christmas present you ever got?*

And my nights weren't much brighter, nasty scenes playing out on the backs of my eyelids: I'm strolling a brightly lit dairy aisle, basket in hand, when one man shoots another in the head at point-blank range, inches from where I stand, blood spattering the white cartons and containers, my face and shirt, my hands. In another, I pull into a diner parking lot and before I get the keys out of the ignition, local toughs pull in on either side of me, so close I can't open my doors. Trapped. Bear maulings, castration scenes, snake fangs to the face, there was no shortage of material. I woke up, sweaty and palpitating, and staggered to the bathroom to brush my tongue and teeth, just the way I would after a bender, as if absolution could be squeezed from a tube of toothpaste.

I sometimes thought of a guy I used to know in college. Mike was a dyed-in-the-wool drunk, farther along than me, fingers never far from a six-pack or shot glass, wouldn't think twice about punching out a friend he'd thought slighted him. But somehow, he quit. I didn't know how or why. I'd just heard he'd gotten married, landed a government job, and had a couple of kids. One day he climbed to the attic of his house and blew his brains out. He'd been ten years sober. I was rattled by the news. I knew I had to make it past ten years. But there was a lot of empty space between now and then, and most of the time I felt like a plastic grocery bag skittering across a desolate parking lot. I filled the void with frozen chicken Kievs, McDonald's, milk shakes, fried chicken, until I felt some solidity return to my life. My doctor had me in for weekly cholesterol checks.

I retightened my laces and started up the pocked lane at a shy, respectful trot, the scratch of the loose gravel underfoot like the static of an old radio song. I couldn't wait for this part of the race. South Street Extension was my ace in the hole, my insurance for breaking 1:50, maybe even 1:45. It almost felt like an unfair advantage, magic beans in my pocket. But first, there were eight miles of rolling asphalt to cover.

I left the dirt and jogged over to the starting area at the northern entrance to the hospital, taking my place among the 350 or so racers, the population of some of the towns I used to cover as a reporter here after college. I didn't recognize anyone, but they all looked familiar. In their faces, I saw forest wardens, soccer moms, college security guards, house painters, state reps. The sun beat back a few gray clouds, and a breeze rippled my shirt and bib number. Within a few minutes, a gunshot tore through the wind and we were blasted to life, a clumpy white monster, all writhing arms and legs.

The crowd splintered as we turned onto South Street, one of the town's most desirable neighborhoods, home to the Middlebury College president, doctors, professors, and lawyers. Columned doorways and wraparound screened porches overlooked wide lawns and leafy trees. Not a shared driveway or rusted car in sight. I was clocking a 7:30-minute pace as

I rolled past the manicured lacrosse and baseball fields, but it felt comfortable. My plan was to keep between 8:00- and 8:30-minute miles the first seven miles so I could kick into sub-eights for the second half of the race. I had it all worked out.

I can feel my heart pumping hard. Not out of fear or panic, but from some different place. Nasty weather swirls in my to-bacco-scarred lungs as I try to move my girth through the water, gulping the cool, chlorine-infused air. More, more. No rest. *I only mean to do four or five laps but swim the entire forty-five minutes of adult swim. Some voice in my head whispers, "Breaks are for pussies," and keeps pressing the trigger each time I touch the lip of the pool edge. When I stagger out of the water, I'm lightheaded and wobbly, a shipwrecked sailor, startling the kids waiting their turn. My legs feel like cement and I can't see straight, the chlorine stinging my eyes, heart kicking like a mule. I shuffle back to the apartment. My fingers shake as I fit the key in the lock. I flop onto the bed and take long, trembling breaths. I stare at the ceiling for an hour, pinned by my own weight, heart still ricocheting. I think about calling Chris to tell her I love her.*

On Main Street, the crowds were thick. The in-town bridge was finally under construction, already spanning a third of Otter Creek. The project to reroute traffic from downtown had been debated for decades, millions spent on engineering and feasibility studies. I wrote about it; every *Addy Indy* reporter in the last twenty years had. Main Street congestion at drive time was as bad as the Mass Pike down a lane, incongruous for this little postcard town with the rolling town green and wooden bandstand. It was weird to see the bridge come to life, like some dusty, lumbering beast awakened from Escher's notepad. Up ahead loomed the Congregational church, sharp white steeple pricking at the slow-moving clouds like a needle searching for a vein. At Triangle Park, the course took a sharp left onto Printer's Alley and fed us into the Marble Works, a commercial complex where I showed up every morning for two years. My first real job after college. Staff writer, *Addison County Independent.* Pa knew the publisher.

My first day the editor sent me to a neighboring town to write about peregrine falcon chicks nesting on a cliff-side. Not exactly the career path of the rock poet I was sure I was destined to be. Worse, I forgot the camera, got lost, and missed the first team of birdmen hiking up. I barely hooked up with a couple of the biologists who'd stayed behind to wait for me. Dressed in tight jeans and my only button-down shirt, I slipped and scrabbled up the rocky trail in Doc Martens with soles worn down to slipper traction. Three times I had to stop the team to rest, my face a ski mask of sweat. I didn't have the strength to swat at the mosquitoes and black flies that swirled above me like a dark halo. I was twenty-four, supposed to be tearing the world apart, infusing the wounds with quotable verse, lighting up the night sky with flares, not caking my shoes with mud and coughing up half a lung in some godforsaken mountain forest in the middle of East Bumfuck. To think most people assumed I was following in Pa's footsteps—and was having trouble at it, no less—made me sick. Yet here I was, peering through a pair of binoculars across a mountain gap. Back at the office, I described the hell out of those baby chicks and their wobbly first steps on the edge of a cliff. My first story. It was even on the front page. I laid the paper across my coffee table and read it several times, my fingertips smeared in printer's ink. I opened my notebook and rolled them on a blank page, a constellation of ovals, smears of me.

We snaked out of the Marble Works and onto Seymour

Street, past the railroad overpass and the wastewater treatment plant. I was running alongside a bearded runner in Carhartt work pants, long-sleeved painter's T-shirt, and canvas ball cap. All he was missing was a tool belt and a stainless steel lunch pail. Costume, Vermont style. I gave him a head nod. He gave me one back. Two dudes moving at the speed of mountain cool. The road curved down toward Otter Creek. Just before the Pulp Mill Bridge stood the first apartment Chris moved into after leaving Shea's dad, a stone's throw from the town dump. Where lead paint lay in peels on the porch, where Chris slept on a pull-out futon so Shea could have a bedroom. Where I visited from grad school on weekends, a paper bag of six-packs in the crook of my arms, oblivious to the hardscrabble, hand-to-mouth life being led there.

I'm sure my drinking pals thought my hooking up with a single working mom was bizarre, even irresponsible. Maybe five-year-old Shea was my first subconscious crack at prying my fingers from the bottle, giving myself permission to separate from old haunts and people. No one could argue with parenthood, even though I didn't know the first thing about daycare or child safety seats. I may have been acting out of character, but Shea wasn't. With her blond hair and chubby cheeks, she thought I was the greatest and I responded. I ran her to McDonald's, tossed her in the warm laundry, and piggybacked her in Lake Dunmore. I made up stories about the Octabutts, misunderstood sea creatures that were half octopus and half buttocks. What that looked like exactly, I wasn't sure—a giant ass with octopus arms? An octopus with butt cheeks for tentacles?—but Shea seemed to be able to picture them and begged for a new tale every night.

Shea loved the tattoo on my shoulder, one I'd gotten when I'd gone cross-country with Jack and Evan when I was twenty years old. It was a duck-billed platypus in a suit, smoking a cigar and carrying a briefcase with the word *Mockba* on it, Moscow in Russian. I'd loved the platypus as a kid, its odd dual nature, and as a drunken adult, who'd lived between two cultures for years, I was vaguely aware of the symbolism. The image was crude and faded and squiggled with flesh-colored lines where I'd picked at the scab. It was designed by a dude I'd met at a party when we stopped in Ohio. I'd wanted to remove it as soon as I took the bandage off. Over the years, I'd gotten a few more tattoos to balance it out and tried not to think about it much. It was a blemish, a reminder. Shea liked to run her little finger over it as the Octabutts circled ever closer.

After the town pool closes for the season, Chris and I join Vermont Sun, one of two gyms in town and the only one with a pool. I've heard the blond-haired, blue-eyed gym owner shaves his muscular legs at the locker-room sink, balls swinging, tanned pecs dancing; that he used to be a picked-on, overweight kid and is now a regional champion triathlete. The locker room is loud and intimidating, and I keep my eyes on the tiled floor. I get in and out as quickly as I can, changing at a corner locker. I stand at the edge of the indoor pool in swim fins and tight thigh-length trunks, encased like a sausage. I can feel the pounds I've put on since quitting smoking. I slide into the lukewarm water, the smell of chlorine deep in my nostrils, the scent of menace. I readjust my earplug and sink beneath the surface. In the pool, I have to learn to breathe again, to synchronize my limbs and lungs. I like muting the world. The water takes me into an almost primordial tunnel of thought. I wonder if three years of sobriety are enough to be recovered. What about five? Sure would be nice to have a cold beer after mowing on a hot summer afternoon. Will these laps reverse the damage like Superman flying

*against the Earth's orbit to turn back time? It's crowded and an
old man with wrinkled skin and a thin beard hops into my lane
and starts cutting through the water. I feel the urge to turn on my
kick. Gonna smoke this dude. Competition, a spark of life. Something I haven't felt in a decade. But the seal on my earplug isn't
tight enough. I can feel the cold tickling my infected inner ear
and I have to stop. I try swimming with my head out of the water
for a few laps, but it's not the same. I look like one of the old ladies with blue hair freshly set.*

*A few days later, ear still aching, I hop on a corner tread-
mill in a back room. I straddle the belt, press Start, and watch
the rubber roll below me, a gentle black current. I build up to a
twenty-minute fast walk, all hips and slicing arms like some kind
of 1980s robot dance. I press the speed button up to 4.9 miles per
hour and my body makes a sound it hasn't in ages—the thump
of a hard heel-to-toe strike. I am running. I'm actually running.
Holy shit. I'm teetering on the edge of something. I picture myself losing my footing, stumbling, and being whipped back like a
dropped towel. I am fleeing, but not running away. In two places
at once.*

The pack stampeded into the opaque light within the covered bridge, sun rays slicing through the cracks and knot-
holes, feet drumming against the wooden planks, before spilling back into the wash of sun and onto Morgan Horse Farm
Road, a rolling, leafy stretch of asphalt, the kind of road mo-
torcyclists cream their leather chaps for. Before long, the first
of two meaty hills poured down. I filled my lungs and charged
up like a bloodthirsty soldier, bayoneting the other runners as
I passed. Despite a headwind, I was hovering around eight-
minute miles. Still on target.

A middle-aged man with sweaty pepper-brown hair blew
by me but stopped some hundred yards ahead, swigging from

his container as he fast-walked. I passed him. He wore knee braces and a fuel belt. Ninety seconds later, he streaked by again, T-shirt billowing, only to slow to a walk. Then twice more. Who was this guy, my Past? I lost him at the turn to Sheep Farm Road.

To the east, farm fields spilled toward the Green Mountains. In the near distance, homes dotted the hillside above Route 7, our old place among them, the shared gravel driveway fronting the busy road, my first house. It was too far away to make out. I slackened my pace on the dirt lane. Until a woman in a red shirt and black butterfly shorts sidled up, matching me stride for stride. I checked the Garmin 205 Chris had given me for my birthday. 9:41. The horror. How could I have let this happen? I brought it back down to 7:55, but the woman didn't budge. We were mirror images. I was getting self-conscious. Dirt turned to pavement and our pace quickened to 7:25. Now, she was trying to shake me. I wondered who she was. A teacher? A trust-fund goat farmer? A man-hater? I was half offended and let dumb pride power-step me through the turn onto Weybridge Street and into a long downhill. There, I let loose, flying tall, hitting 6:43 and holding it, an insane pace for me. But she was behind me now. That'd teach her. Before I knew it, I was running uphill. I galloped to the top, keeping my pace under 8:20. My lungs were bellowing like an accordion on speed, my throat raising a hand for water. As I crested, I lost a half-minute on my pace. Nothing to worry about, plenty in the bank, I assured myself.

But as I turned onto Freeman Way toward the college campus, it was as if someone had ripped the key from the ignition. My thighs turned to lead and I could hardly shuffle. My Garmin flashed 10:03, 10:21, then 10:38. I was shocked. My pace was swan-diving off a building, hurtling toward the pavement below. 11:03, 11:12, 11:26. Where was the bottom? My legs were sucked into the earth, quads sputtering to a halt. I felt lost, terrified, out of control. Outside my body. Had I really hit the wall at mile 6? This was beyond the bacon feeling. This was the whole pig. My Garmin might as well have read: *Say your prayers, Asshole.*

Then it hit me like a boot to the groin. Over the past three years, I'd been tearing around the Charles River, paths flat as paper, whittling down my speed, chest puffing up at my sub-7:30s. I hadn't run a single training hill for this race, hadn't even looked at the course map. Why bother? I'd lived here for years. But could I really have been so dumb? This was Vermont we were talking about. All that speed work didn't mean shit here. That was just flash. Vermont's landscape was all about patience and persistence and quiet strength. It wouldn't even wipe its ass with flash. I'd literally lived up to the flatlander name. I'd committed the sin of forgetting and now I was paying the price. My bank of early 7:20s had been chewed up. The woman in red was long gone. The pepper-haired man with the knee braces chugged by me. I looked away.

I find the upstairs treadmills taken. That leaves me with the main gym downstairs, a sprawling, glass-fronted space with mirrored walls where guys are bench-pressing, lifting free weights, working the speed bag, and hauling ass on a bank of high-tech treadmills. A living mural for all the gym traffic to take in and a place I avoid. There might as well be a sign that reads FIT MEMBERS ONLY. *Only one machine is free, between two women in jog bras and thin shorts, their feet a blur, eyes in the zone. I step onto the cold belt, fiddle with the controls a few seconds, pretending I know what I'm doing. I decide to try "Golf Course," a half-hour computer-generated route. I set the maximum speed at 5.2 miles per hour and start walking the three-minute warm-up.*

I'm a little nervous. I've never run a full thirty minutes without stopping. I catch sight of myself in the mirror, pale hands busying themselves with my Discman.

Five minutes in and my shins start aching, sounding off like a gathering mob. I turn my attention to the figure in the mirror running at me with quiet determination. The pale arms, the pumping legs, the bobbing head. Is he running for help? For his life? An inner saboteur spends several minutes trying to convince me I've not eaten enough, that I'm feeling faint, that guys like me never go the distance. The argument has merit. I'm already melting like a candle. The treadmill's computer speeds me up and down, then up again. I'm tethered to the back of a sadist's pickup. One foot in front of the other has become a monumental task. I'm ready to start making deals.

In the mirror, I notice the dawn starting to gather in the windows behind me, and the faster I run, the lighter it seems to get. The speed of the whitening light is in sync with my footfalls. I begin to make out the parking lot, the traffic along Exchange Street, the treeline. I feel like a Peruvian villager powering the community's only television, my feet moving the belt that slowly raises the sun. I wish I could hop off and pause the day at this half-light, stop everyone, and everything, at this moment in time. Inspect them, inspect myself, figure out my place here. A series of beeps snaps me back to the task at hand and I'm walking in cool-down mode. I wipe my brow. I've made it. Hold on, something has happened. I'd moved somewhere beyond my conscious mind, dug beneath the prison wall; there was a whole world out there, fields of energy. All while moving in place. I think I've just tapped into some kind of magic. The women beside me are gone.

I tried to control my breathing and relax my mind, choosing not to react to my jellied quads, the burn in my lungs. I put my mind in my shoes. I was sure there was energy in the ground beneath my Brookses and I tried to draw it out like a fever. The path sloped past the gray stone campus chapel, a couple of dormitories, and toward the arts center, and just when I thought I'd need to drop out of the race, miraculously, I felt my legs coming back around, degree by fat degree. By mile 7, my pace had dipped back down to 9:17, then 8:34. My body had somehow reset, my confidence rebooting. Man, that was fuckin' scary. To lose control so suddenly and completely felt like a car crash, and to have all the crushed metal and broken bones and bloodstained tires reassembled out of nowhere felt like divine intervention.

Before I knew it, I was jogging up South Street, past the college track, where Chris and I sometimes ran laps on the weekends, along with groups of old ladies walking, the white lines keeping us all in place. From there, I could see Porter Hospital where the race had started and beyond that the rough tan ribbon of South Street Extension that lay a quarter-mile away. I leaned toward the dirt, my dirt, struck by the vast distance contained within the short span between the oval track and the gravel road, the travel possible without really going anywhere.

Leaving the college track, I bump into my landlord walking toward Main Street, his white T-shirt soaked through with sweat and tucked into his canvas shorts. Walter is an accountant, tall and lanky, a lifelong runner. "The best piece of running advice I can give you is buy the best shoes you can afford," he says, his face flushed with color. "That's all that's between you and the road." Then Walter mentions a road he likes to run, out past the college track, beyond the hospital, all dirt. "Very pretty, quiet," he says. "Easy on the knees."

The following weekend, I park at the hospital and walk to the end of South Street, which turns to gravel after a few hun-

dred feet. I'm still wearing my backless bargain sneakers, not quite ready to shell out the big bucks, not sure I dare call myself a runner. The dusty lane is pocked, strewn with small rocks, but no one else is around. It rises and keeps rising, a long, deceptive incline. My lungs feel like two sticks being rubbed together by a survivalist. Then the road evens out and curves past a field and a couple of houses set back behind the trees. I settle down and straighten my back. An old golden retriever lies on its side in the driveway next to a recycling bin, its soupy eyes watching me as I pass. It's quiet.

I can hear my feet scratching and picking, a new sound. Before I know it, I'm jogging downhill, past an abandoned cow barn and rusty grain silo, the wind in my ears. This is a whole different kind of running, no help from a machine. It's almost exhilarating. But within minutes, I'm trudging up another hill, the smell of manure choking my nostrils. At the top stands a tall barn, several slats missing like knocked-out teeth. A pack of muddy farm dogs darts from the house across the way. I clench and drop my hands close to my groin, the first place I figure they'll lunge. I picture myself being picked apart while I lie panting, face-down on the dusty road. The mutts dart up, bellies, snouts low to the ground, and run alongside a couple of paces, then fall back. Jesus, that was close. When I get back to my car in the hospital parking lot, I reset the odometer and drive the route. I've covered 4.2 miles. Good lord. It feels like the Fourth of July.

Finally, I came to my beloved seam and crossed over. I was back in the cradle. Surely my old run would balance the ship, puff out the sails, and steer me toward a sub-1:50. I thought of the sublime runs here. I could still play them in my head like movie scenes: a 4:30 a.m. run lit up by moonlight or greeted with a sunrise so red it's as if the sky had been slit by a scalpel or a pack of coy dogs following me behind the mist. The road crunched underfoot as I mounted the rise, a few empty cans and a pizza box resting in the gully. Orange-tipped stakes dotted the first field, and a large sign showed an architectural rendering of a collection of townhouses. I hadn't noticed that the first time.

I could feel the temperature shift. The tire that used to hang on the field entrance had been removed. In the distance, the Green Mountains held down the scene like a paperweight. I reached the college recycling center at the top of the hill. A couple of ladies sat against the large car gate, applauding. Within a few minutes, a motorcycle rumbled toward me and in its wake the men's leader. He was alone. I mustered some energy and clapped for him, hoping to feed on some of his winning vibe. I felt his breeze. He didn't break focus.

I eased back as I made my way down that first long downhill, not wanting to repeat my quad-crushing mistake from mile 6. When I got past the grain silos and to the flats, I settled into an 8:30 pace before the short climb up to the Eddy horse farm. Just four miles to go. I spied a couple of mares in a far field, grazing, sipping from the earth. The missing slats in the side of the gray barn had been replaced with fresh blond boards. Just beyond the troughs, two folding tables sagged with paper cups of water. I grabbed one and the liquid went down like a guillotine, barely coating my throat. My pace had slowed to 9:47. By my calculations, I could still make 1:49. But just barely.

I shuffle past the hospital and nursing home, down the cracked asphalt toward the gravel road. It's 4:30 a.m. and black as pitch. All I can see are the whites of my running shoes, flashing like eyes. I follow their every move. Above me, the sky is a velvet bowl pricked with starlight. I push into the darkness, my body hardly apparent, as if running into a dream.

My shoes slap and scrape against the tight earth. I'm intimate with this gravel, the ruts and runnels, the sink of the shoulder. I can make out the silhouettes of farmhouses, tractors, and roadside vegetation. I know where to step to avoid a divot or patch of loose rock, but still a certain faith is required for each footfall. Who knows what fresh hazards lurk in the dark? Fallen crab apples, an unearthed stone, roadkill. I am running through church.

Suddenly, headlights angle up from around a bend, creeping, feeling for the corner before straightening into a blinding force. For a moment I am stunned, pinned in place, at mercy, waiting for the shotgun blast or bottle against my skull. I force myself to the side of the road, stumbling into oblivion. I picture myself lit up like a television angel. A pickup truck crunches by. I am spared.

I regain my footing and realign myself with the road, arms slowly sawing, feeling for the notch. Down a hill, through the flats, then a slow climb where I'm met with the warm smell of horses, mixed with the scent of hay and turning apples. Sweet as incense. The hilltop farm presides over a small swell of valley. I can't see into the paddocks but I feel the eyes on me. Proud, shiny haunches, cavernous nostrils, manes combed with speed and freedom. What I'm pushing my body to do, horses were made for.

With every step, I'm conversing with my surroundings—an exchange composed not of words but of urgent breaths, of flesh against earth, flesh against flesh, of fluids. Wind forces tears from my eyes, my nose is running, sweat glazes my brow and spreads through my shirt. I'm turning myself inside out, blurring the lines between me and the universe. I am consuming, and being consumed by, my landscape. I gulp the air, then expel myself back into the world, to nourish some other entity—a plant, a horse, a farmhand.

Fence rails pass me by, then sheds and barns, the peaked roofs arrowed toward the sky as if in homage to the Green Mountains in the distance. The wood is supple, softened by rain and time, but proud as any church steeple. The crossroads rush to meet me. I turn right, down the swamp road, past the small dairy farm.

In the swamp, trees creak and groan. Water has crept over the road in places. Suddenly a wet noise, close by, in my ears. Wings beating water? Claws? I sprint to the other side of the dented path, clear my throat, and growl into the darkness, an attempt to scare off whatever's there. Silence.

An overgrown logging road marks the end of the swamp and telephone poles emerge on my left, impossibly tall crosses, cables sagging like smiles against the slate-gray sky. When I reach the electrical transformer a few hundred yards down the road, I turn back, my feet flicking bits of fresh gravel.

A pale blue has gathered at the edge of the velvet bowl. I lick the sweat from my lips as the irrevocable slide begins. The sheet of darkness is being pulled back, revealing by degrees fields of yellow-haired corn, lawns planted with political signs, a broken flower pot, tire-flattened frogs. Pinks and grays smear the sky. A train whistle aches in the distance. I run my last mile fast as if to beat the sun, the seconds scattering from me like spores. When I reach the seam between dirt and pavement, I let go. The motion drains out of me like blood as I slow into a brisk walk, sweaty and exalted, my gait transformed, spirit and body lined up like stars. And the day has just begun.

I followed the barbed-wire fence with my eyes, the wild grass lapping at the knobby posts. Overhead, the tree canopy broke open. The sky was scabbed over with clouds. I dashed around a couple of divots, my feet scraping at the dust. Watching the runners on their way back, one in a Boba Fett mask and cape pushing a baby stroller, I wondered about the normal, happy lives I assumed they led, the fun they were apparently having, the pitfalls they'd avoided. I started feeling heavy, dragging, alone. I cast my mind backward.

Every one of my apologies had been drafted on this road, at six miles an hour. My sweat the ink, my lungs pushing the words up into my brain. It had taken me five years without the bottle to realize that not drinking was hardly enough; it wasn't the same as slapping a nicotine patch on your shoulder. Things could never be right until I reached out to those lives I'd injected with turmoil, the ones I'd left behind. My MO had always been: Torch connections and don't look back at the smoking mess. A Russian goodbye. On the surface, that felt stoic, but it was, in fact, cowardly and ignoble. To say "I don't see you anymore" was a gut stick. I'd broken hearts to feel empowered, breaking a bit of my own every time, and leaving a hardness lingering in its place.

I started with Chris, apologizing for my moody weather, for always adding a six-pack to a grocery list she could barely afford, for never going to bed until the last beer was drained, for drinking around her young daughter. I had confessed to Shea, who'd had no idea even though when we first met I was rarely without a bottle in my hand and the Octabutts had crawled out of my beer-soaked mind. I apologized to my parents for the lies, for the canceled visits, for driving their car while drunk, for their sleepless nights, for making them see my hands shake any time I tried to write anything in front of them.

I reached out to old girlfriends, who'd borne the brunt of my recklessness, my selfishness, my anger. I'd used these women to feel OK about myself, to feel confident, but when we got too comfortable, I broke up with them or forced them to leave me. I judged them harshly for being with someone I loathed, mistreated them to impress my friends. It was on this dirt stretch that one day I realized how long I'd been terri-fied of being in my own head, a loner who couldn't bear to be alone.

These women were now scattered around the country, the world. Some responded to my e-mails, some didn't. One ar-ranged to meet with me, then canceled the day before. I was grateful for every mailing address I received. But I hadn't been sure how to begin this kind of correspondence. The offenses took place years ago — some were blacked out, many misted over by time. I wanted to explain who I thought I was at that time and some of the why as best I understood, to take re-sponsibility for my words and deeds, and let them know that despite my harsh talk, obnoxiousness, and scummy actions, I had cared for them. That I was sorry for how I made them feel. That there had been smiles. That I saw them.

When I got back from my runs on South Street Extension, I wrote down notes and the snatches of half-conjured mem-ories, the names of people I'd slighted, the way I needed to phrase things. I penned every letter by hand, standing at my dresser. I wanted a physical act, to feel my legs tire, my hand cramp, my lower back ache. I wanted to trace a line from my heart to my brain to my hand to the page. I wrote their par-ents and siblings, too, families who had embraced me, who had hung Christmas stockings with my name on them. I wanted them all to hold what I held, to see the imprints of my palm on the page, to have something more lasting than words on breath, my Braille. To make it tactile felt poetic, almost ro-mantic, but somewhere inside me the letters felt unsatisfying, as if I were apologizing to a piece of paper. It didn't feel pain-ful enough.

I concluded by offering to tell them all these things face-to-face, to hear any comments they might have. None of them took me up, which both relieved me and bummed me out. I felt like they were the ones who had moved on without looking back, with husbands, children, interesting careers in cool cities, and I was the one stuck in the past. Maybe I'd held less import in their lives than I'd imagined; they'd dispensed with me as easily as I had with them. More than once, I had to remind myself that this process wasn't about me.

Navigating this old road again, I wondered if I'd said enough, put down the right words. Not everyone had written back. Was it too late to ask for an eyeball-to-eyeball meeting or had I blown my chance? Would a sit-down really seal the deal? I still fretted over how I'd handled some of the letters. Had I put more effort into my prose style than the sincerity of my sentiments? Was part of me trying to make them miss me a little? Maybe drunks never find complete closure and a certain joylessness and low-fevered regret is part of the transaction, that hoping for happiness is somehow selfish. But that seemed to go against the meaning of our place on this Earth, of being alive. Sometimes, even sober, it's too easy to choose to be a ghost.

A runner streaked by, one arm coated in tattoos like a graffitied rail car. I thought of the platypus on my shoulder, the one Shea loved so much, even faded and clouded. After years of forgetting it was there, ignoring that part of my body, I'd begun thinking about it again. The feeling of regret hadn't necessarily disappeared but it had a different edge, of hope somehow. I touched my shoulder, feeling the small bump of scar tissue that had grown within the blur of green and blue. I thought of Casey, how I'd mishandled that apology, even got a bit snarky after she changed her mind about dinner, mad that she would deny me this important piece of work. That circle was not closed. Maybe it never would be. But it was

OK. It was OK to have loose ends, to feel the sting of regret. The key was not to tear myself apart for feeling this way. The pain wasn't a stamp that I'd done wrong but a reminder that I know I can do better. Almost twenty years had passed since that ink needle first broke my skin, five years before Shea was even born. I was now, finally, coming around to her point of view.

As I saw my sub-1:50 start to dissipate, the thought of walking seeped through the keyhole in my mind, under the door. Running was always a reality check. I told myself to roll with it, that with all the people everywhere and the bright sunlight, this wasn't my old run. In one way, though, South Street Extension was doing what it always did—teach. Things don't always work the way you planned. Maybe you never outrun your demons, but if you maintain forward motion you might just get them to tire a little. Two miles left.

At last, I reached the turnaround, the railroad crossing and cow barn a short jaunt ahead. I circled around the orange road cones and tilted back toward the rusted grain silo in the distance, the undulating carpet of fields. Seeing all the runners behind me lifted me up. Amazing what feeling superior can do for you. I grabbed a slice of orange from a spectator and tore it apart with my teeth, the sweet pulp dripping down my chin, my fingers sticky. I turned back and whispered thanks, to nothing in particular.

I pushed hard up the final hill, no longer aware of the other runners. Seven minutes later, I hit the finish-line mats. I'd left my sub-1:50 back on South Street Extension, along with some other things, but I posted an official personal record. I was pleased. I had tried. I had pushed through the tough spots, something never reflected in the net time columns. I felt drained and restored at the same time. As I chugged a second bottle of water and looked for Chris, a friend of mine from the bad old days found me, his infant son strapped to his back. Ian had lived in Burlington at the same time as I did and was now an admissions officer at Middlebury College. He was a smart, good-looking guy, quick with a joke or a story. He never got too drunk, or maybe I just couldn't tell. I remember arguing a lot with his girlfriend. "I couldn't believe that was you out there," he said. When I told him I was revisiting my sinning grounds in a pair of sneakers and a bib number, he quipped, "A different kind of bib, eh?" I smiled and wiped the sweat from my eyebrows. A perfectly fine point. I couldn't have put it better myself.

        1 hour, 51 minutes, 36 seconds (personal record)
              Average pace: 8:31-minute mile
             133rd place out of 354 finishers

Washington, DC

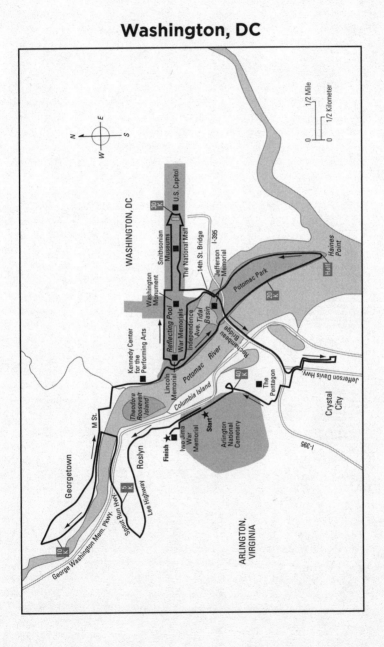

7

35th Marine Corps Marathon

Sunday, October 31, 2010

T HE SUN WAS STILL bedded down. I followed the throngs of silhouettes, not sure where we were going. It turned out to be a long, dark walk from the Pentagon City subway station to the Athletes' Village. Across the inky channel of water spread Lady Bird Johnson Memorial Park and beyond that, the Potomac River. On the opposite shore, the Jefferson Memorial glowed. Farther north stood the Washington Monument with its single red eye, like a bloody tip, stabbing at the sky. I darted down to the water. The riverbank foliage was perfect cover. Of all my marathons, this had to be the bladder-friendliest. I thought about going a couple more times just for the hell of it.

The ghostly outlines of planes banked overhead from Ronald Reagan National Airport as we shuffled past the fortress walls of the Pentagon. When my family lived here in the

1970s, it was called Washington National. That was a long time ago: an almost forgotten era of Kiss, *Star Wars,* the Iranian hostage crisis, the undefeated Miami Dolphins, the Atlanta child murders, and Steve Martin doing the funky King Tut. My feet grew cold in the early-morning chill, so I stopped at a bench and changed from my slides into running shoes. Laces tied, it was now officially on. I rejoined the line.

We'd been told to expect beefed-up security for this race. A few days earlier, a gunman had taken potshots at the Pentagon, the National Museum of the Marine Corps, and an empty recruiting station. The shootings took place when no one was around. The guy was attacking symbols of the culture. The police said they believed the man was a disgruntled former marine. It seemed plausible to the authorities that the Marine Corps Marathon might be in his gun sights, too. (The shooter later turned out to be a young, homegrown Muslim radical acting alone, who had videotaped himself in the act.)

Even though Washington, DC, wasn't one of my sinning grounds, I'd put the Marine Corps on my list, too. It felt right to end my marathon streak by going back to the beginning. I was born in this city and began formation here. It had always felt like a button on Pause. Plus, the Marine Corps Marathon was my father's race. I remembered meeting him at one finish with my mom. Pa was all crumpled under a foil blanket like a half-eaten gyro, worn out, vulnerable. The sight didn't make sense to my young eyes. Pa was an upright man, always moving or sitting straight. Now, he was still, sprawled, glowing. I had no idea what he'd just gone through or why, but he looked soft as if I could reach out and touch him. Later, I told him I wanted to run the next one and even ran with him on a few weekend mornings. I never made it to the start line; I can't remember why. A couple of helicopters buzzed overhead, chopping up the memory, and I pinned the bib number to my shorts.

The Marine Corps is the fourth-largest marathon in the country after New York, Chicago, and Boston. It doesn't cough up prize money, only bragging rights, and some thirty thousand prideful runners had signed up. An American man has won five of the past thirty-five contests, and an American female has taken the women's division every year but six. The race was nicknamed the People's Marathon, a Soviet-style moniker if there ever was one. In fact, Washington's hide wasn't so different from Moscow's. The memorials, the parks, the roads, all named after figures out of the history books—the past expressed in concrete and marble. The need to mark transcends political ideologies, borders, and ages. Otherwise, we'd crash about meaninglessly, blind in the woods.

To counter antimilitary sentiment in the wake of the Vietnam war, a marine colonel named Jim Fowler started the Marine Corps Marathon. He pitched the contest to his superiors as a way to promote goodwill, serve as a recruiting tool, and give marines a chance to qualify for Boston. In 1976, some 1,175 sets of running shoes gathered at the inaugural start line, including a pair strapped onto the feet of a skinny, bespectacled military reject named Nicholas Daniloff.

A national security correspondent for the now-defunct United Press International, Pa first laced up for the Marine Corps when he was forty-one, one year older than me. There were no timing chips (you got a wooden stick at the finish line with your place marked on it), no energy gels, no expos teeming with booths and fit and trim salesmen with the solution to your every running woe, even ones you didn't know you had. Pa wrote an essay for the *Washington Post Sunday*

Magazine, back when newspaper editors saw marathoning as a lark. I read it on the plane from Boston. Even Pa, a committed man if there ever was one, wasn't concerned with finishing, just getting fifteen miles or so. "I tried to prepare by running a minimum of eight miles a day and more on weekends," he wrote. "I never did keep to my training plan. Too many things got in the way. But I ran one hour and forty-seven minutes once. I think that was about 13 miles."

It'd been weird to read this sustained piece of inner-Pa, a voice I didn't quite recognize, a lost recording. I sometimes thought my father didn't have inner thoughts, or they were so cold and remote they might as well have dwelled at the bottom of Lake Baikal. He was made of facts and column inches, veins running with printer's ink. After being turned down by Navy Officer Candidate School because of high blood pressure, Pa had worked his way up the journalism ladder from copy boy to foreign correspondent. He'd wanted to draw a paycheck from the State Department as a diplomat, not cover it as a reporter. The rejection by the military still stung, reinforcing a feeling of inferiority perhaps encoded in the family genes. Passing buff marines on the roads became silent revenge, safe little thrill kills.

Wasps have taken up under the eaves of the front porch. I'm scared of their segmented bodies, their dangly legs. Circling around their gray papier-mâché cone, they might as well be wearing tiny leather jackets with pentagrams on the back. It's summer; I'm ten years old. One afternoon while I'm playing in the front yard, several wasps land in my hair, which is long and unbrushed. They get caught in the tangles. And the stinging begins, one hot plunge after another. Tiny switchblades. I start swatting at my head. Pa hears my wails and dashes out. I'm in hysterics. I run in circles, trying to get away. They keep stinging, furious, buzzing, my scalp burning over and over with fresh pain. I run and run, but I can't run from my own head. Pa finally grabs me by the shoulders and gets me to stop. Lungs heaving. He tries picking them out, getting stung himself, too. "Oww, goddamn it!" He brushes hard at my hair. I stomp my feet, howling, my chest drenched, my shirt front balled in his fist as I try to wriggle away. But I've stopped crying, as if I've run out of pain. I study Pa's face, the focus in his eyes. Clean-shaven. Hard Adam's apple. Sandpaper skin. I don't remember how long it takes, but he finally gets them all out, brushing down my hair, his fingers running over my scalp.

At the approach to the Athletes' Village, clusters of marines, clad in desert camos, were manning tables, devouring the dark with bright flashlights, checking bags and jackets. But it was so crowded and void of light, you could have easily slipped past with a machete or hidden a gun in your waistband. The appearance of security was apparently as effective as the real thing. That illusion can be as powerful as reality, was as exhilarating as it was unsettling.

The crowd shimmied toward a massive parking lot that glowed with large white tents and was ringed by a fleet of brown UPS trucks, which would haul our gear to the finish area. Inspirational slogans like "Beat the Bridge!" and "Remember Iwo Jima!" flashed on a giant LCD screen. BTO's "Takin' Care of Business" throbbed through mega-speakers. Groups of runners mingled, snapped pics, shared Gatorade, admired each other's costumes. The LCD screen flashed times for multidenominational prayer services. I took a seat on a curb beneath a tree and began unpacking my bag.

As I watched a runner spread a towel in the Pentagon parking lot and go through vinyasa, the sun salutation, biceps and calves encased in black compression sleeves, I thought of this paragraph Pa wrote in his marathon essay: "Somehow the last-minute preparations of others always upset me. To the south of the Iwo Jima Memorial—the starting and finishing line—several confident young men are pushing their shoulders against spindly trees in some kind of exotic warm-up exercise. Several young boys, who look about ten years old, are proudly prancing about announcing, 'We're going all the way.' Couples in sweatsuits are walking arm-in-arm as if this were some kind of semiannual picnic which they wouldn't miss for the world, instead of what I expect will be a grueling foot race . . ."

Reading that the first time had almost shocked me. I felt the same way before every race. *What do they know that I don't? Shit, how prepared am I?* I was always jumpy as hell and needing to piss. Pa even wrote about that, too, feeling like a "puppy needing to leak" as soon as the gun went off. Within those sentences, I felt some distance between us close, his breath closer, like seeing him in that finisher's foil again. It was running, not breaking stories and speaking truth to power, that could bind us. Sitting on the cold curb at the edge of the parking lot, I wished I'd brought a bottle of water, another banana, extra goo packets. But there was no turning back, not after months of training, the 20-mile runs, the black toenails, the chafed nipples. I reminded myself this was just a 26.2-mile victory lap. No choice but to embrace my nerves, to treat my fear as room for improvisation. Pa's thirty-five-year-old paragraph wasn't enough to convince me that we're destined to become our parents, but I couldn't deny that genetics and environment conspire on some level. And was being like your mother or father inherently such a bad thing?

"It smells like pong in here," Mom says as she balls up my sheets and gathers them in her sturdy arms. "Open the window."

Is that edge in her voice? Is she mad? I look at the hard translucent plastic undersheet left behind. It looks like gray ice. I hate seeing it. I hate the crunchy sound it makes beneath me. I don't have friends into my room in case they sit on my bed and ask what that sound is. In sixth grade and still wetting the bed like a baby. I strip out of my wet pajamas and trudge to the bathroom and wipe down my legs and stomach with a washcloth before school, but I can always still smell me, smell the pong. Mom had promised me a prize if I can go a week without wetting my bed. We have a calendar set up and for each dry night, she pastes in one of those gold star stickers like at school. I've been fantasizing about a new baseball mitt. But my dry streak ends at five. It's not fair and there's no one to complain to, no one to explain. I stare at the calendar boxes, the rows of three, of two, all the blank boxes. I'd almost made it this time. Now, I have to start all over again.

The MCM starting line had shifted over the years, but the race still ends at the Iwo Jima Memorial statue, perhaps the most iconic image of World War II: six GIs working together to plant an American flag on the Japanese hilltop after a long and bloody battle. Today, the crowds of adrenaline-juiced runners would toe the start line between the Pentagon and Arlington National Cemetery, placing participants squarely between those who serve and those who have sacrificed. Even the least patriotic couldn't help but feel stirred by the young runners with bib numbers pinned above their prosthetic legs, or the fathers and sisters wearing T-shirts with iron-on portrait shots of young marines in dress uniform, framed by a pair of dates. A reminder that somewhere, not here, blood was spilling, limbs were being torn from bodies, realities were being shattered.

This year, the MCM was also marking the 2,500th anniversary of the Battle of Marathon, an epic clash between the Greeks and Persians in 490 B.C., and a torch lit from the eternal flame at the Tomb of the Athenians, commemorating those killed in action, was burning at the starting line a quarter-mile away from the staging area in the Pentagon parking lot. The victory at Marathon is credited with saving democracy, but most people associate it with the birth of the 26.2-mile distance race that bears its name. As is true of most creation myths, the waters are muddy. The popular story goes like this: After their unlikely rout of the Persians, Greek generals at Marathon dispatched a runner-messenger named Pheidippides to Athens, twenty-five miles away, with the news that the enemy was now sailing for the undefended capital. The story climaxes when the panting, sweat-drenched Pheidippides reaches Athens and shouts, "We were victorious, rejoice!" then drops dead of exhaustion.

According to the Greek historian Herodotus, an Athenian runner named Pheidippides was indeed sent to run with a message, but it was from Athens to Sparta. He charged some 140 miles over hilly, rocky terrain, arriving the next day, begging the Spartans for help at Marathon. But religious laws kept the warriors sidelined until the next full moon. So an exhausted Pheidippides turned around and hoofed it back. Over time, the two journeys were baked together and came out of the oven as piping-hot inspiration, laying the foundation for the marathon debut at the first Olympics in Athens in 1896. A falsehood that gave birth to a reality and a movement, a way of life.

Announcements blared over the loudspeakers for runners to make their way to the start. I followed a thick, slow-moving crowd and we all found ourselves on the wrong side of the barricades. With just a couple of minutes left, I squeezed through an opening in the gates at the 4:00–4:39 corral. Again, I was shooting for a 3:59:59. But this time, I had a secret weapon. An MIT and Harvard researcher named Ben Rapoport. Ben had figured out a scientific calculation, based on your weight, VO2 max, and pace, that would tell you exactly how many energy-radiating carbohydrates you'd need to consume without hitting the dreaded wall, that excruciating state which results from a depletion of energy stores in the muscles. The wall forces the body to slow down dramatically as the legs simply run out of gas. Paces plummet, runners walk, hearts break. For many middle-of-the-pack runners, it happens after mile 18, about two-and-a-half to three hours in. The wall is every marathoner's greatest fear. My greatest fear. I'd hit some serious bricks and mortar in both Moscow and New York. If I could neutralize that monster, maybe I'd finally break four hours, reach the other side, at last.

Ben was a twenty-nine-year-old, sub-3:00 marathoner and had made a tradition of lecturing Harvard Medical School students on endurance physiology an hour after he crossed the Boston finish line. He decided to share his findings with the wider world, and the debut story on his formula was cued up for the next issue of the Public Library of Science *Computational Biology* journal.

In it, he explained that the body produces energy aerobically in two ways: through fat metabolism and the breakdown of glycogen and glucose, in both cases facilitated by oxygen. During intense exercise that approaches one's VO2 max (the maximum capacity of an individual's body to transport and use oxygen), most of the energy comes from glycogen, which is stored in the liver and muscles, mostly in the legs. A typical untrained person on an average diet can store about 380 grams of glycogen, or 1,500 kilocalories, in the body. Intense running usually consumes more than 800 kilocalories per hour, draining the supply within two hours. But a 155-pound man needs about 2,900 calories to finish a marathon. Hello, wall. But if the same runner maximizes his carbohydrate intake, he'll have 650 calories in the liver and 2,270 in the leg muscles, for a total of 2,920 calories. No wall.

The key was a steady pace, and Ben's calculations were based on running at a consistent speed. If you run too hard or accelerate too often, your body selectively chooses carbohydrates as fuel and you risk depletion. But if you slow down, your body can burn more fat, sparing some of the glycogen. A lot of marathoners know about carbo-loading to ratchet up their stores, but it often involves guesswork and a lot of pasta dishes with plenty of hidden energy-wasting oils. Ben was offering a guarantee.

Our first conversation at an Au Bon Pain near the MIT campus was intoxicating as he painted a scenario where I could safely run a 3:30 if I wanted. "But let's start with a 3:59," he urged. Ben was thin, wore glasses, and had short brown hair and had already run twenty-one miles that morning in preparation for the New York City Marathon later that month. He was focused and accurate, all business, not an ounce of fat on his thoughts. When I told him I was psyched to be one of his guinea pigs, he corrected me, saying, "You'll just be proving an established fact."

I had my VO_2 max measured and Ben told me to draw up a meal plan, detailing my breakfast, lunch, dinner, and snacks. He had me revise it several times until the amount of my daily carbohydrates lined up with the pace I would need to run to break four hours: 9:07. I kept the paper in my pocket, folding and unfolding it like a treasure map. Over the four days prior to the race, I'd stuffed my face with pasta, bagels, rice burritos, large pretzels, and bananas. I counted every protein gram and fat and carbohydrate calorie. When I topped my daily allotment, my heart glowed. At the time, I didn't really understand the science, how carbs convert into energy, why glycogen is stored in your liver, how the body chooses to burn energy. I just knew that promise and possibility lay in those numbers.

What I didn't tell Ben was that I'd done a shadow calculation for a sub-3:50, a minute faster than Pa's first MCM. However, being stuck in the 4:00–4:30 corral was not part of the plan. I should be with the 3:30 folks; they might all overtake me, but at least they wouldn't get in the way and I could maintain momentum. But it was too late to jump ahead. I heard a muffled boom—a Howitzer cannon, appropriately—quickly papered over by cheers. Bodies decked out in bright fluorescent shorts and singlets and jog bras poured out like spilled jelly beans down Route 110 toward Lee Highway.

Once we got going, the muffled pounding of shoes on the asphalt sounded like a relentless surf. I spied my shadow on the roadway just a few paces ahead of me, a black, plodding shape. It seemed to move so slowly, yet always stayed a few steps ahead. A string of clouds passed overhead and the figure vanished. I carved and squeezed forward where I could, raising an apologetic hand whenever I brushed against someone.

After twenty minutes, we were rolling down Spout Run to the George Washington Parkway. Pouring down to meet the road on either side were lush, sloping hills covered in vines and ivy. I remembered this terrain, the thick foliage, the clay, the veins of creek water, the crayfish, the caves, the dappled sunlight, the southern landscape that colored parts of DC, that separator of North and South. Rock Creek Park, Wolf

Trap, Palisades, the names came flooding back. This was my Once Upon a Time.

I hold the sharpened pencil beneath Thomas Armstrong's butt as he sits in his chair. He lets out a wild scream, the broken No. 2 still stuck in his backside. Mrs. Fox, our second-grade teacher, rushes Thomas to the bathroom where the nurse pulls down his pants and tries to remove the lead tip from his buttock. The next day, Mom marches me over to the Armstrongs' to apologize. Thomas won't see me, but I tell his mom I'm real sorry and give her a get-well card I made. I have no idea why I've done what I've done. I don't know why I do a lot of things: why I eat dry cat food, why I eat gum off the sidewalk, why I painted a white stripe on Mandy's cat, why I still suck my fingers. When we get back, Mom tells me to go upstairs and take my pants down. I do as I'm told, never once thinking of running away. I listen to arguing downstairs. Then footsteps. I'm frozen, holding my breath, my backside cold. Pa walks in with a wooden spoon. I can't look at his face.

A mile later, we were streaming across Key Bridge (named after anthem writer Francis Scott Key) and into the capital. Running along the C & O Canal, I noticed a large woman in spandex pants and loose shirt was jogging on the dirt path across the narrow waterway. I was almost jealous. She was simply running. We were racing. She didn't seem to notice us and there were a lot of us not to see. I wanted to cheer for her for some reason. If I'd come across a race in my early running days, I'd have been intimidated enough to turn around and go home and watch TV. Then the leaders came into view and I turned away. The pack was slender, focused, wrapped in thin, wispy shorts and singlets, like tossed rolls of ribbon, already on their way back from circling the Georgetown Reservoir, some four miles ahead of us. They chatted easily with each other.

Those first eight miles, there was quite a bit of talk among us middle-of-the-packers, bits and pieces of conversations bubbling all around me, merging into one long hum. It reminded me of a sweat lodge ceremony I once attended in the woods of Vermont for a story. I'd heard some Native American groups were incorporating *inipi* into rehab programs for Indian inmates in other parts of the country. But in the squat earthen room dug into a private hillside it was mostly white men with ponytails and names like Spirit Bear and Big Sky Speaks Thunder huddled around a pit of glowing stones. Still, I wasn't the only former addict sitting naked in the dark, dripping with sweat, hot lungs heaving, tongue a broiled sponge. During one round, the twenty or so participants were invited to speak their prayers aloud at the same time, all of us creating a cacophonous cauldron of open-hearted energy in the pitch black; no senses, no forms, just lightning bugs in the mind of a god. Now, seven years later, all around me those glowing bugs again.

As the road hugged the water, I knew that over the years, running had infused a necessary spiritual weight into my life. While I loved the Bible stories and Greek myths my mom used to read to me, I'd been raised without religion, my formative years spent in a decidedly atheist country, and embraced a cynical and morally dubious young adulthood. Mom later lamented not sending me and my sister to church. But running had cracked me open, letting light into the hard-to-reach corners. It was a confessional, baptism by sweat. You can't be false when your legs are screaming, your heart pounding, mouth gaping. You feel naked, and when you feel naked, you feel naked in front of something. Perhaps God or some cosmic energy or simply the wonder of nature, but something bigger than yourself, stirring a need to honor, to prostrate. Running was my daily prayer service, the marathon my vision quest.

A mile later, I gained on three soldiers in green camo pants and rucksacks. As I moved closer, the outlines of one of their bags came into focus. I saw what appeared to be an artificial leg sticking out of one sack. I glanced at another soldier. Another leg. The third runner was packing a plastic arm. As I passed, I saw the men were surrounding a hand cyclist, who was cranking the wheel with his one good arm. The rest of his limbs had been lopped off at the joints. Sweat was pouring down his cheeks, his focus locked. I gave him a wave, then felt stupid because obviously he couldn't wave back. I patted my heart a couple of times in a gesture of camaraderie. I knew that even if I lost use of my legs, my arms, my eyes, I could still "run." Being a runner was a state of mind, something you evolve into, and once you arrive, you never go back.

Before I knew it, we were rolling down M Street in Georgetown. We passed over the canal and the towpath that runs along the water. As a kid, I used to call it the "toad path," which always made it seem like a magical place, something out of a Beatrix Potter tale. I once threw all my toy guns into the canal, pleasing my pacifist father no end. Could they still be in there somewhere, had they become part of the silt? Up ahead, a traffic light turned from green to yellow. Even though the roads were closed to traffic, I instinctively sped up to beat the red light, heart swooning, mind racing, thinking about cops lurking like alligators. I was struck by how certain things become fixed in our minds, society's rules and norms, man-made constructs, almost hard-wired into our brains and hearts. It's scary when you realize how malleable we all are. Like the body expecting motion when stepping on a dead escalator or learning to look in the opposite direction when crossing a street in London. But by the same token, we should be able to get unused to anything, too, no matter how calcified it once was.

I keep sniffing until I can't get any more air through my nostrils. My lungs, chest, blown up like a balloon. I need to draw in the entire world. To push beyond bursting, to fill every corner. Each sniff louder and deeper. More, more, and then a little bit more. I can't help myself. I must do this until it can't be done anymore. Until I have inhaled myself through my nose. I don't know when this tic started or why. At the Kennedy Center one evening with my parents, I sniff through the entire performance, snorting up the music, the chandeliers, the opera boxes, the dancers on stage. The good-looking, well-heeled couple in front of me keeps turning around to stare. I have no idea why. They must think I'm famous. I pretend I don't notice them and keep on sniffing. Mom and Pa nickname me "Sniffovitch."

I drew even with an older runner with a buzz cut, running on a Cheetah blade and swinging a prosthetic arm. An able-bodied runner pulled up and asked who he was running for. The man thought for a second, then answered, "I'm running for me today. This is my first marathon." Another runner quipped, "Yo yoan then." We all looked at him. Then he explained: "You're on your own." I liked that. *Yo yoan.* I turned the words over on my tongue until they became a single sound, a rhythmic chant. *Yo yoan, yo yoan, yo yoan.*

Who was *I* running for? Not just for small kids with big ears and a bed-wetting problem. Not just any kid whose body conspires against him, who feels on the outside looking in, desperate to control some element of his life, who burns with shame. It went beyond me. As my drinking grew into my sidecar and then my twin, I'd become fascinated with former child stars like Danny Bonaduce, Dana Plato, and Jan Michael Vincent, their post-career drunken descents splashed on the covers of the tabloids. They had it all, then they didn't. When I got sober, this *People* magazine rapport evolved into feeling for the average Drunken Joe, too—a driver who mowed down a young mother in the crosswalk during a blackout, a wasted student athlete who woke up to find himself charged with attempted rape, the drug-addled teen driver who'd paralyzed his girlfriend taking a turn too fast. Lives changed forever in a single moment; such moments had been coiled within me, too, but for some reason never sprang. I know these people had little or no memory of their actions, the pathway gone, just skid marks and a broken guardrail and the cold rush of empty air. They were totally lost and sunk deep in the horror of spending the rest of their lives with themselves and a crime they couldn't trace. I could condemn their actions, but I just couldn't condemn them as people. Their names are burned into my brain. Whenever I read of another incident—and there's always another incident—I was filled with that breathless feeling of being pulled back from a ledge. Could this have been me, with just a few more cosmic tweaks, a couple more fateful incidents? I told myself no, but I knew nothing was absolute. Goddamn, I'd been lucky.

If I couldn't feel for the lost ones, their detritus-strewn lives once the waves pulled back, then this had all been a pointless exercise. I'd be able to catch my twin, catch the boy, but would have nothing to say to him. A red military transport helicopter thrummed overhead, filling the air with a giant buzzing heartbeat.

At the approach to Hains Point, the halfway mark, the

path narrowed and the crowd tightened, with maybe five or six runners across. There was no room to pass. I tried sniffing out even the slightest opening. I just had to be patient. My pace climbed to 9:20, then 9:40. I spied a curb and hopped on, tippy-toeing down the narrow expanse as if on a tightrope. I brought my speed down to 8:20 and held it there for a half a mile. I came upon another hand cycle, another cluster of companion runners; a wounded officer with his men, it turned out. Each of them wore a shirt that read PAIN IS JUST WEAKNESS LEAVING THE BODY. Normally, I found a lot of running mottoes bumper-sticker-like and trite, but this one I liked.

I lost time at the next water station, the road clogged with stopped thirsty runners, all elbows and sweat and no motion. I had never allowed myself respite in past races, even if that meant wearing half my Gatorade on my chest. I couldn't say I'd actually run a marathon if I stopped to walk while I drank, more like 26.19 miles. I'd become a slave to accuracy—and haunted by inaccuracy. Probably why I hated being a journalist. Every mistake crushed me, every angry phone call or e-mail a devastation.

Pa is reading a book on his bed. I'm lying next to him. It's Sunday morning. I'm six years old. Quietly, stealthily, I dig my fingers into his armpits. "Yow!" he yells, leaping and tossing his book aside. "Coochie coochie coo!" I scream and dig my fingers in deeper. He has his hands under my arms. I'm in hysterics. Zeus is romping and barking. "Coochie coochie coo!" Pa screams in a falsetto voice and dives back at me. I'm laughing so hard, it almost hurts. He finally stops tickling. I roll over, smiling, trying to catch my breath. I love those words, this feeling. I can hear him breathing, too.

I slowed to a brisk walk, grabbed a cup, and came to a halt, coating my throat with control, relishing the relief. I wasn't bothered, surprised that my fists weren't clenched, a curse word not loaded in my tongue. I crushed the cup, tossed it in a bag held by a marine in baggy camos, and moved on. I started up more easily than I'd thought. There was no pain. I wasn't dependent on my own momentum, only on my own permission. I had the power to stop and then start again. Something had shifted inside, something accepted by my tired, ghostly captors, who just wanted to go back to their own families, their own way of life. Eight miles and one bridge to go.

From Potomac Park, we strode past the Jefferson Memorial before entering the National Mall and running by the Lincoln, FDR, Korean War, and Vietnam Veterans memorials, and all the Smithsonian museums, including my childhood favorite Air and Space Museum, where every visit was capped off with a bag of freeze-dried ice cream. Spectators were wandering the course and had to be avoided. I thought about Pa's description of his body at this same point in the race: "Great from the waist up, terrible from the waist down." I was starting to feel terrible all over. And it was just mile 19. Must be all that weakness getting the hell out, I told myself.

We passed the Tidal Basin and continued along Jefferson Drive before turning onto the endless 14th Street Bridge that crosses the Potomac back to Virginia. At first I took the slogan "Beat the Bridge" to mean conquering the long, late-stage incline at mile 20. A battle cry. But it actually referred to reaching the crossing before 1:00 p.m. when it reopened for traffic. Stragglers—those running a fourteen-plus-minute pace—were to be bused to the finish line. I already felt for those poor bastards.

The bridge's long, wide span felt as big as an ocean. I

couldn't see the other side. I throated down my first packet of energy goo, mint chocolate flavor. It tasted awful, my mouth sticky as if I were wearing caramel lipstick. I felt like I was on a treadmill; nothing was getting closer. A dude in a cow costume was handing out Halloween candy from a plastic jack-o'-lantern at the halfway point. I gave him a wave. At last, I began to feel the bridge sloping under my feet back toward land.

But by mile 23, I could no longer ignore the exhaustion settling in, pulling me toward earth. I was running through wet concrete, quads filling with rocks. I pulled the empty water bottles from my belt and chucked them one by one, like sandbags from a doomed craft. If I could have snapped off my head, I would have tossed that, too. I cursed the iPhone sunk in my pocket like a brick. I choked down three more packets of goo. With considerable effort, I was able to dial my pace down to 8:53 for a spell and then keep it at 9:15, which told me I hadn't hit the wall, that my body was still obeying my commands. But to break four hours now, I needed to run nine-minute miles for the rest of the race, my regular morning run pace. But I knew what my legs were telling me: *That ain't going to happen, my friend.*

At mile 25, we wound down an on-ramp and into a stiff wind for the last 1.2 miles. A few minutes later, I looked at my Garmin and watched the relentless display turn from 3:59:59 to 4:00. First thought: race ruined. I'd disappointed Ben, who was following me online. Maybe I didn't carbo-load precisely enough. We'd never talked about hydration or navigating crowded water stops or starting in the right corral. Maybe my once-alcoholic liver had already been compromised, a poor specimen. I was never good at math, a bad student. I heard the words *was never good. Bad.* But they came from my mind, from a discarded script, not my heart. I felt no surge of negative emotions, just the urge to keep moving forward.

I looked at the Garmin display again just to watch it hit 4:01, then 4:02, to see my goal slip farther away, like a tongue feeling for a freshly pulled tooth. I looked away, again expecting disappointment to slow me down. But I felt fine, my pace intact. I was surprised at how easily I accepted the reading. I felt myself loosen. As a breeze crossed my face, I recognized something. It was the numbers. The fucking numbers. They had become my master. That black part of my brain was still crafty, no doubt. Over the past year, I'd let myself be choked by numbers—carbs, pounds, fat grams, miles, minutes, calories, bib numbers. The sub-four had become not only a barrier to break, but a tool with which to judge myself. The digits on my Garmin, the same thing. The self-criticism that drinking first softened, but ultimately exacerbated, still echoed somewhere inside me. It was here in DC that I first started judging myself, feeling shame at being me, too young to realize the far-reaching implications. And it was here where I could snuff them out for good.

I realized the sub-four was not literal. It was artificial, a smokescreen my mind had thrown up. That number really didn't mean anything. It was what it stood for. A symbolic barrier. And I realized I'd already broken it. Every time I struck out in a nor'easter or a subzero freeze or jet-lagged after an international flight or after being up half the night battling bad Mexican food. The worse the weather smashing against my windows, the more I was drawn to lace up. In the dark, howling wind and cold, lashing rain, I heard both a bully's taunt and a Siren's song. There was no way I was pulling the covers over my head. Getting myself out there, taking that first step, saying yes. I'd already been doing this for years, I realized, overcoming, taking all comers. I also understood there had been an accumulation. It was OK for me to rest, too. There was no one to answer to anymore but me.

In many ways, recovery is simply a synonym for life. We are all just trying to survive being born. Recovery, like life, like death, is ultimately an individual journey, a single-seat rowboat with no life preserver. If you capsize, you have to learn how to turn your own jeans into a flotation device. There's always work left to do, more road, more corners, more hills. It is the grappling with invisible monsters on a road with no name that leaves fang holes on your neck. The struggle is in learning to open your eyes, to see.

Up on the hill I finally spied the red finishers' arches. Even though it was a short climb, I staggered and shuffled, my body all but broken down. I didn't even see Chris leaning over the barricade, clapping her heart out, hooting and hollering. When I crested, it was a short jog to the finish. It looked a thousand yards away, but I felt like I'd already crossed over.

The lawn is thick with people but not crowded. It's the Fourth of July, 1976. The Bicentennial. I'm seven years old. Blankets are scattered about, weighted down with picnic baskets and families. I run around barefoot, excited about the fireworks. I'm streaking across the grass in denim cutoffs and striped T-shirt. My hands are sharp, each finger pressed together to form a knife, slicing through the humid air. The wind combs my hair. I can hear it in my ears. When I get back, Mom and Pa are leaning against each other, Pa wearing jeans and sandals. My mom is in a floral skirt. And a floppy hat? Mandy is off somewhere. It's getting dark but I can still see forms and outlines. I flop down with my parents and lie on my back. Mom brushes my sweaty bangs from my eyes. I hear a rumble. I stare up at the dusky sky, the Washington Mon-

ument rising into infinity. The other week at school, we'd gone on a field trip that ended with a visit to the very top. The people below looked like ants. Now, I was one of the ants. A huge explosion cracked the air, and reds and oranges and greens crashed against the sky, all of us lit up like electric ghosts, the monument throwing off muted colors. Warmth everywhere, light weeping down. I could have stared forever.

I crossed just ahead of a woman dressed as Minnie Mouse. I smiled at the symbolism, unafraid to embrace it. I had beaten out my childhood. I thought of the other symbolism that had leaked into my life: that my feet had grown almost a full shoe size since I started running; the sweet sustenance in the form of pie from NMH; boarding a plane for Moscow on my sober anniversary. Just as I had evolved from runner into marathoner, the symbols had morphed, too—into symptoms. All around me, I saw symptoms of my recovery. But, really, my childhood and youth were never there to be recovered from or beaten or even completed. I'd given them a power they didn't need. Victory wasn't the answer. The past just needed a shelf, someplace to be still, to be seen if need be. I was the answer. I was always the next chapter. It had always been me who stood in the way of perfection.

As I shuffled to a walk, I stared at the Iwo Jima Memorial, the exhausted bronzed men draped over each other like a single twelve-legged creature, the collective effort to raise and plant a single flag. To claim territory. I'd never seen the statue before, despite growing up here. The image seemed appropriate, the struggle, the triumph of that critical World War II battle. Less at stake today, of course, but it was all about honoring what had come before, what it had taken to get here.

A marine sergeant placed a medal around my neck. I thought about saluting but could barely raise my arms. I grabbed a black Mylar cape and wrapped myself up, quickly borne along by the walking crowd, no idea where we were going, or where I would meet Chris, just grateful at last to be part of the current, to be going somewhere. I touched my fingers to my forehead and saluted my maker, my flawed source, and forgave us all his creation.

4 hours, 3 minutes, 34 seconds (personal record)
Average pace: 9:17-minute miles
5,089th place out of 21,873 finishers

Epilogue

Arsenal Bridge Route

June 2011

The rain is light, kicking up a dusty smell from the sidewalk that mixes with the brackish musk seeping off the Charles River. A few shells are plowing through the water, followed by a coaches' launch. The light is fading. I take to the bald dirt path that hugs the boulevard, softening the throb in my left knee and right glute. A steady stream of cars headed home after work whizzes by me, windows up, wipers lazily swiping. I never pictured myself an evening runner. The thought of changing into shorts after eight hours at my desk, several meals coating my belly, the day's baggage clinging to me like a rucksack, seemed onerous. But worst of all, sleeping on that shaken snow globe of endorphins seemed like such a waste. Plus, I'd always prided myself on hitting the streets before most coffeemakers in the neighborhood started gurgling, all light and fresh, perfectly equipped to conquer the slumbering world, a surprise attack, infusing my body, my day, with fresh-dewed energy. Running at night, I just never got it.

But a few weeks earlier, I'd developed sharp lower-back pain, a knife tip deep in my glute that made even walking aggravating, a toothache in my ass. And not wanting to be left out of the number, both knees began chiming in, too. The pain was intense in the morning. I had to grip the handrail going down the stairs like an old man. I worried that all those marathons had been a mistake. It was like a Greek tragedy: Chasing my story had sped up my deterioration and hobbled the one thing that kept my head above water. Not even an anguishing two weeks off made much difference. I just felt fat, sluggish, guilty, and in pain.

But I couldn't not run. The ache yielded somewhat as the day went on, so one evening, after dinner, I'd popped three Advils, tugged compression sleeves over my calves, and hobbled into action, grimacing through the first mile, trying to run taller and engage my core to take the strain off my back. Still, my lower back sang like a tortured canary and took days to quiet down. I'd run through colds and fevers and shoulder and groin strain, but this pain wasn't budging, so I finally went to a physical therapist, who told me the muscles supporting my lower spine weren't firing.

This wasn't good, as I'd just started training for the Chicago Marathon. I know, another marathon. Doesn't make sense. I had no connection to the Windy City, had been there only a couple of times to visit my sister at college. But I wanted a race experience that was free of my past, where I was running 26.2 miles for its own sake. No sub-four, no Garmin, my wrist naked. No numbers in my ears, only the wind. Forget Vibrams; this was minimalist running. My cop running pal, Bill Coffey, and his writer wife, Brigid, offered to put me and Chris up. Bill volunteered at the mile 19 fluid station every year and said he'd run with me for the last 7 miles. Normally, I'd have turned down such an of-

fer. Running is such a solitary experience for me; I can only have conversations with myself, in my head. By mile 19, I'm usually pretty spent. Having to form words and audible thoughts was scary. But I told Bill that that'd be great, that I'd look forward to seeing his eager, earnest face. Friends with a cop and former prison guard; who'd a thunk it? Life is funny. I couldn't wait to see what would come next.

Running in the evenings feels different. Things are winding down but there's a thicker, crowded, louder energy, as if the world isn't yours alone; everyone's claimed a piece and you're just looking for a free outlet to plug into. Runners, bikers, Rollerbladers, dog walkers, bench sitters, geese feeders, hand-holders, Frisbee tossers, panhandlers, they're all out there. There's more traffic, endless centipedes of hard-eyed Boston drivers, and more running shoes paused at streetlights, more waiting at the mercy of the blinking orange hand. Seeing the white "Walk" figure is the most beautiful work of art, a shadow box with the raw power of a cave drawing. Go, be, become.

It isn't long before I'm again reminded that even within the simple act of running, a seemingly monotonous activity, there can be complex, radical change. Whereas in the morning run there is an infusion of energy, in the evening there is a peaceful draining of the accumulated thoughts and stresses that can keep me awake. I embrace the sunset, the inevitable end, grateful for another day above ground. The orange "Don't Walk" hand allows me a respite from the back pain, a chance to catch my breath and look around, not to be locked into my head. To watch a mother goose lead her goslings to the water, a child running with his arms out like airplane wings. It's good practice for interruptions, to embrace the breaks in the routine, the fissures in my mind that may always be there. I'm running in a state of gratitude and acceptance.

The drizzle turns to rain, the thick clouds eating up the lingering light. I watch the eight-oared shells as I scrape along the pavement, the unison of their blades squaring and feathering. My back still barking a little. In between the breaks in the riverbank foliage, I sometimes used to see Pa out here in the mornings. He often logs nine miles up and down the river before the sun fully spreads, and he competes in the Head of the Charles, the country's premier rowing regatta. Now in his mid-seventies, he seems to row more than ever. Mom sometimes gets irritated at Pa's obsession and complains that he's denying his own mortality. I sometimes picture Pa out there being followed by a shell manned by a tall figure in black robes with scythes for oars. I think he'd be thrilled to check out with his hands wrapped around oars, rowing forever into the black. But since I began running in the evenings, I don't see him on the water anymore.

Just below the Arsenal Street Bridge, I pass a large woman in capris, windbreaker, canvas ball cap, and running shoes walking in the opposite direction. When I turn around at the bridge, she is ahead of me a ways. At first, I don't notice she's running because she's moving so slowly. Then she stops and walks again. As I get closer, she runs again for a spell. My heart softens. She's just starting to run. This is the beginning. I feel blessed in a way, as if I've spotted a rare act in the animal kingdom. The becoming. A foal with splayed legs, covered in afterbirth. Her cheeks are infused red, ripe, a couple of sweat beads clinging to her temples.

I can tell she's trying not to breathe so hard, her lips tight. I'm dressed in my pitiless sunglasses, fluorescent yellow MCM shirt with the black Iron Eagle logo, wispy shorts, and compression sleeves on my calves. An orc chasing a hobbit. Watching her weight sway side to side, the plodding steps, I'm reminded that the medals, the race T-shirts, the PRs, the Big Four marathons, all of that didn't matter a bit. It's still about the basics, one foot forward, then the next, running a bit farther each time, the wind against your cheeks. She's beautiful. I'm almost jealous; she has yet to discover that running in the pouring rain allows you to own the weather, the day, everything around you. Or the first time you inexplicably burst into song along with your run mix, that moment when running feels like dancing. Or the moment when you forget that you are running, the realization that you are not a prisoner of your own mind. That you can achieve not only presence but a state beyond, one devoid of expectations, fears, criticisms, of self.

She can sense me behind her shoulder and gives me a scared, almost intimidated, sideways glance, as if I've caught her doing something she shouldn't or doesn't have the right to do. There are hardness and fright in her dark eyes, shadowed by the brim of her cap. I smile and give her a wave, holding it an extra beat, and keep going, momentarily moving through her bubble. I make sure not to cut back into her path too soon. A few moments later, I look back. She's still running. My heart lifts. For that moment, she's making it. I move around the bend. I'm filled with love. I can't feel the ground, only air. A perfect run.

Gratitude

For my indefatigable and enthusiastic agent, Wendy Sherman, who always returns my calls and e-mails and gracefully shepherded me through the mysterious and sometimes painfully slow-moving world of publishing. For my astute editor at Houghton Mifflin Harcourt, Susan Canavan, who had the faith and courage to take on this project when it still looked like an unmade bed. She read my drafts with obvious care and challenged me to get at the deeper story. She taught me more about narrative than I ever expected and unquestionably contributed to my growth as a writer. I'd also like to extend my appreciation to the entire HMH team for helping bring this book to life, from design to copyediting to publicity.

For Zeeb Green, a Revolutionary War soldier buried in Brandon, Vermont, who influenced my life from beyond the grave. A 2005 radio commentary I wrote about Zeeb prompted an e-mail from his great-, great-, great-granddaughter Mary

Anne Coffey, a seventy-year-old genealogy buff living outside Chicago. We became pen pals. In 2009, Mary Anne put me in touch with her son, William Coffey, who was also tackling his first Boston Marathon that year. We met, and Will's then-girl-friend Brigid Pasulka, a talented and soon-to-be-award-winning novelist, later showed my proposal to her agent, Wendy Sherman. Both Will (aka Will/Bill) and Brigid were enthusiastic supporters and generous readers of the manuscript. Rest in peace, Mary Anne. I'm sorry you didn't get a chance to read this, but you are, in no small part, responsible for its existence.

To my first writing instructor at the University of Vermont, Toby Fulwiler, who saw himself in me, hiding in the back row. Also from UVM, Jean Kiedaisch, a persistent supporter.

To all my nonfiction writing classmates at Columbia University's Graduate School of the Arts, Class of '99, especially the Lis Harris workshoppers. And my champions among the faculty, in particular Lis Harris, Patty O'Toole, and Richard Locke, all of whom helped me believe that someday I could do this thing.

For Charles Burch, who taught more than he ever knew and left before I had a chance to thank him.

To the folks at *Runner's World* who first published my essays on running, snatches of which appear here.

For my bosses at Boston University's Department of Marketing and Communications, who gave me time and flexibility to work on this book. To all the OGs at *BU Today* and *Bostonia* (RIP, 10 Lenox) and especially Nelia Ponte, Mary Cohen, and Bari Walsh for their insightful feedback and support in the early stages. Thanks also to fellow scribe and sculler Amy Gutman for her prompt and helpful feedback.

To my parents, Nick and Ruth Daniloff, who took me

on a hell of a ride and ultimately gave me something worthwhile to write about. And my sister, Miranda (you'll always be Mandy to me), who always reached out even though our paths diverged far too early. To cousin Claire Goodman, who always radiated belief. To my English grandmother, Baba, who, when I was still a fourth-grader struggling with penmanship, predicted jail or writing for me. Only time will tell, I suppose.

And for my wife, Chris, who has always been my first and last reader on the page and in life. Her influence on me is too big to measure, her contributions to this book too numerous to count. She lives and breathes in these pages and by now knows them by heart. Thanks for admiring my leather jacket all those years ago, Babe. I'm so grateful it fit, and even more grateful that it still does.

To my daughter, Shea, whose wit and grit have shown me the true meaning of strength and perseverance. So honored to be yours, Bubs. Keep on shining on.

And to everyone who appears in this book, thanks for coming into my orbit and making life all the more interesting.